Optimal Supply Chain Management in Oil, Gas and Power Generation

Optimal Supply Chain Management in Oil, Gas and Power Generation

Editor

Bhushan Kulkarni

scitus
academics

Optimal Supply Chain Management in Oil, Gas and Power Generation

Edited by **Bhushan Kulkarni**

Printed in 2017

ISBN: 978-1-68117-428-0

Library of Congress Control Number: 2015936542

Contents

Preface

This book discussed on a toolbox for large-scale capital expenditure decision making and for transforming capital and operation expenditures to exert a visible financial impact in oil, gas, and power companies. Drilling and production was concentrated over only a few geographical areas, and refining was done at small-scale refineries. In the 1930s in Saudi Arabia, workers crushed minerals by hand to create drilling mud. The California-Arabian Standard Oil Company initially employed Bedouin tribes to guard their fields and supply lines, transitioning as its operations expanded to government police and eventually to private security. In 1938, before the advent of domestic refining capacity and regional pipelines, Saudi Arabia exported crude by barge to Bahrain. By the 1940s, tanker trucks were transporting oil from the production site to refineries. Each held 40-SO barrels, compared to 120-215 barrels for tanker trucks today. Pipelines were also used, but they were not the main mode of transportation.

Editor

A Review of the Wood Pellet Value Chain, Modern Value/ Supply Chain Management Approaches, and Value/ Supply Chain Models

Natalie M. Hughes, Chander Shahi, and Reino Pulkki

Faculty of Natural Resources Management, Lakehead University, 955 Oliver Road, Thunder Bay, ON, Canada P7B 5E1

ABSTRACT

We reviewed 153 peer-reviewed sources to provide identification of modern supply chain management techniques and exploration of supply chain modeling, to offer decision support to managers. Ultimately, the review is intended to assist member-companies of supply chains, mainly producers, improve their current management approaches, by directing them to studies that may be suitable for direct application to their supply chains and value chains for improved efficiency and profitability. We found that information on supply chain management and modeling techniques in general is available. However, few Canadian-based published studies exist regarding a demand-driven modeling approach to value/supply chain management for wood pellet production. Only three papers were found specifically on wood pellet value chain analysis. We propose that more studies should be carried out on the value chain of wood pellet manufacturing, as well as demand-driven management and modeling approaches with improved demand forecasting methods.

INTRODUCTION

In a time of great uncertainty and drastic change in the global forestry industry, many companies have found it necessary to shift away from manufacturing conventional forest products and refocus their attention on value-added forest products as well as managing waste (wood) more efficiently. Specifically, the creation of renewable fuel sources for the production of energy, such as wood pellets, has become very popular in recent years [1, 2]. Wood pellets have many advantages, including high density and heat value and low moisture content, and are relatively convenient to transport and store [3, 4]. Wood pellets are used for both residential and industrial purposes for the production of heat and/or electricity. There has been an increase in global demand for wood pellets and Canada has responded to this increase by exporting large volumes of wood pellets overseas [5]. A number of wood pellet production plants are emerging globally, thereby creating more competition. Canada is currently among the top producers and exporters of wood pellets [6, 7] but due to this increased competition, Canadian manufacturers must find ways to stay competitive in the global market. One way this competitive edge can be achieved is by optimizing production and logistics within the value chain [8].

This paper provides a review of the literature surrounding wood pellet production, the value chain, supply chain, and wood pellet market analysis. Specifically, the four objectives of this paper are (i) to review wood pellet characteristics and production, (ii) to describe and examine the (wood pellet) value chain and the supply chain, (iii) to summarize the broad scope of varying types of value chain and supply chain mathematical modeling approaches, and (iv) to offer our perspective on the gaps in the literature within these three objectives as avenues for potential future research. These four broad objectives contain multiple subobjectives as well. Under objective (i), the components of wood pellet production and the most valuable characteristics of wood pellets will be examined, including modernized certification standards. Objective (ii) will contain a comparison and examination of the benefits and hurdles of demand-driven value chain and supply chain management approaches. The importance of demand forecasting to combat uncertainty and enhance leanness and agility within the value chain and supply chain will be evaluated, and a synopsis of the wood pellet market will also be included under this objective. Objective (iii) will highlight the number of specific value chain studies found within the various mathematical modeling approaches.

Value Chain is a concept introduced by Porter [9] that describes a chain of key activities performed within an organization that generates value relating to a product (or service). The value chain tracks the activities required to bring a product (or service) from its conception to fruition in terms of the value that is added to the product (or service) as it moves through the supply chain, which consists of the set of processes required for its completion and delivery [9]. The value chain serves to create an understanding of how, where, and how much of the value created by the product is achieved at various product refinement stages throughout the supply chain. The premise is that each activity along the value chain will create value that exceeds the cost of providing the product (or service), therefore resulting in net profit for the company [10–13]. The goal of value chain optimization is to maximize the value achieved at each stage throughout the supply chain, while minimizing costs. The value chain, even though it is based on internal operations, also has connections with suppliers and retailers, and competition between any of them will damage the entire chain [14]. Porter [9] also emphasized the importance of cost reduction and/or reconfiguration of the value chain in order to obtain a competitive advantage in the marketplace. Value chains differ dramatically, based on the type of product being produced, and no single chain may be used to satisfy one industry [14]. Sathre and Gustavsson [15] emphasized that linking product processes and byproducts provides a beneficial approach for individual firms to add value and increase profit.

A set of firms or a linkage of separate agents, each with their own individual value chains that pass materials forward and bring products or services to the market, is called a supply chain [9]. During this review, it became apparent that there is some ambiguity about the concept of the value chain versus the supply chain. Many of the articles and reports we reviewed offered no differentiation between the two chains and in many cases used the two terms interchangeably. However, Mentzer et al. [16] sorted through the multitudes of varying definitions to provide a more cohesive view of the supply chain and defined supply chain "as a set of three or more entities (organizations or individuals) directly involved in the upstream and downstream flows of products, services, finances, and/or information from a source to a customer." Despite this definition, confusion unfortunately still exists as evidenced by this review.

The value chain and, therefore, the supply chain can become both more productive and profitable if companies focus more of their attention on total supply chain costs instead of just parts of the supply chain in order to optimize performance and revenue [17–21]. Value chain optimization involves coordination between a (manufacturing) firm's various nodes, of the supply chain, through appropriate value chain governance at the operational level, which will allow for the overall supply chain to become more efficient as well [8]. Wood pellet manufacturers and other industry stakeholders need a precise understanding about distribution channels, sustainability, long-term forecasting, and methods to improve their operations within the wood pellet supply chain, to ultimately improve their value chain. Different operational management methods of the supply chain need to be identified for improvement, and the exploration of different modeling techniques will help in determining the best fit for wood pellet supply chain modeling under changing (market) conditions. In this paper, our examination of the peer-reviewed literature available to date provides this identification and exploration through a summary of the existence, and merit, of modern supply chain management techniques, while also gathering information on modeling techniques to support managerial decision-making. This summary allows member-companies of supply chains, mainly producers, to recognize shortcomings of their current management approaches and provides an excellent starting point from which an in-depth analysis of specific management techniques may be executed. Implementation of techniques most conducive to achieving improved efficiency and profitability of the operations of specific companies, and their supply chains, is the long-term goal of this review paper.

We have arranged this paper as follows. The methods section describes and presents the number of papers we reviewed in an organized fashion, relating to each objective. The results and discussion section contains subsections for each objective and explains the number of papers we found for objectives (i) to (iii), relative to the total number of papers we reviewed. We discuss the main points of the reviewed papers and how they relate to each objective and subobjective in the subsections of the results and discussion section. Under the sub-section for objective (IV) we highlight the literature gaps we discovered while searching for peer-reviewed studies relating to objectives (i) to (iii) and provide recommendations for future studies to fill these gaps. We conclude the paper with a recap of the main findings of this review and the benefit of filling the literature gaps we have highlighted.

METHODS

This section describes the number of papers we reviewed and how we catalogued them under objectives and sub objectives. We examined a total of 153 sources and categorized them relative to the first three objectives of this research paper (Figure 1). We reviewed a total of 23 journal articles and reports related to wood pellet production, characteristics, and certification in order to achieve objective (i) (Figures 1 and 2). We researched value chain and supply chain management perspectives through examination of 68 peer-reviewed publications to fulfill objective (ii) (Figures 1 and 3). For execution of objective (iii) we reviewed 62 peer-reviewed publications in which models were developed to improve either value chain or supply chain efficiency (Figures 1 and 4). We met objective (iv) through the realization of objectives (i) to (iii).

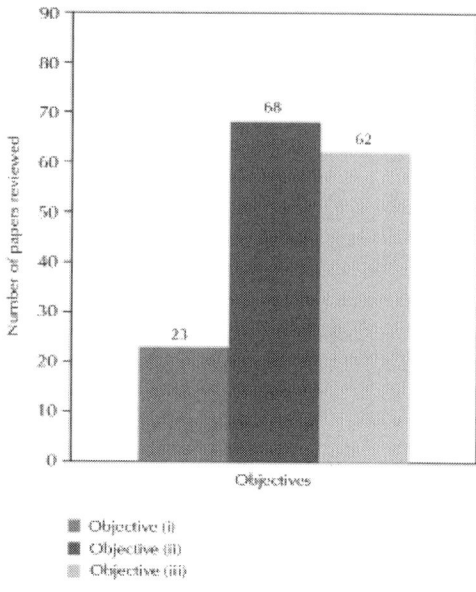

Figure 1: Total publications reviewed for objectives (i), (ii), and (iii).

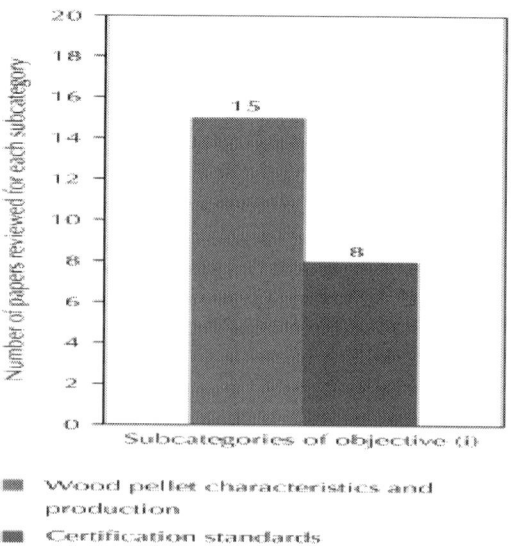

Figure 2: Publications reviewed for the subcategories of objective (i).

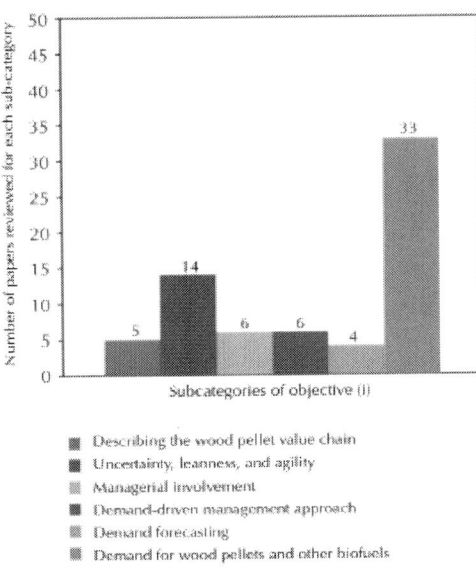

Figure 3: Publications reviewed for the subcategories of objective (ii).

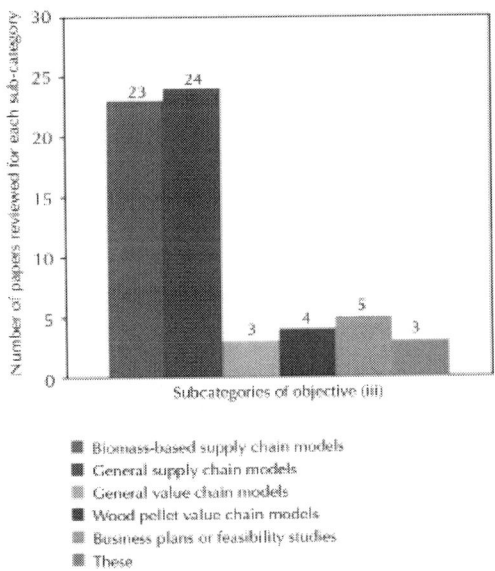

Figure 4: Publications reviewed for the subcategories of objective (iii).

RESULTS AND DISCUSSION

In this section, findings from our literature search relating to each specific objective are discussed in terms of the relative number of studies we found concerning each objective. The information extracted from these studies is summarized in order to determine the literature gaps that may be filled by future research.

Objective (i): Wood Pellet Production and Characteristics

Of the 153 papers reviewed, we found only 15 (10%) describing the production process of wood pellets and/or explained their beneficial characteristics, of which only three (2%) gave an in-depth analysis [3, 22, 23], one of which was a Canadian-based study [23].

The basic steps in the pellet production process (from raw materials to pellets) include (i) drying, (ii) grinding, (iii) conditioning, (iv) pelletizing, (v) screening for fine separation, and (vi) packaging/storing of final product [22, 24–27]. Raw materials for pelletization in many countries worldwide are mainly wood shavings and sawdust from the wood processing industry [25, 26, 28]. To create one tonne of pellets with moisture content between 7 and 10%, an approximate volume of $7.5\,m^3$ of sawdust must be processed (at moisture content of 50%) [1, 3]. Once formed and cooled, pellets are either filled automatically into small (i.e., 40 pound) bags for residential consumers or large bags (i.e., 650 kilograms) for larger customers or stored in bulk in silos or halls [3, 22]. Raw materials cost and (when using wet raw materials) drying costs comprise the majority of total pellet production expenses [28–31]. As pellet-plant size decreases, production cost increases [32, 33].

Densification of wood pellets, as a result of compaction, allows for greater homogeneity of the product, enhanced combustion efficiency, and efficient transport and storage [30, 34–36]. A study by Pa et al. [37] indicates that the combustion of wood pellets requires less primary energy than the combustion of undensified wood waste and that pellets emit lower levels of harmful emissions (i.e., carbon monoxide, nitric oxide, and particulate matter) than wood waste. Sultana and Kumar [38] used PROMTHEE (Preference Ranking Organization Method for Enrichment and Evaluation) to determine that wood pellets were superior to pellets made of other feedstocks, namely, straw, switchgrass, alfalfa, and poultry litter. This method was used considering 11 criteria, both quantitative and qualitative, under three differently weighted scenarios for use in large-scale heat and power generation plants. The results indicate that wood pellets were the "best source of energy" for all scenarios. Naik et al.'s [39] study also found specifically that Canadian pinewood had the best physicochemical characteristics and lowest detrimental emission levels as compared with other biomass samples.

Wood pellets are used for small-scale/residential systems, district heating, and cofiring with coal in large-scale power plants [5, 25, 34]. District heating refers to a network-bound heating plant that is centrally located and connected to a number of buildings (i.e., a residential "district" comprised of households or schools, smaller businesses, etc.) [3]. In North America, wood pellets are most commonly used in small-scale/residential heating systems, and modern versions of these small-scale systems have become automated to the point that they require only a minor amount of maintenance [3, 30]. High standards for pellet fuel quality are required in the residential sector, with a high level of homogeneity required to achieve fully automated operation and complete combustion in small-scale furnaces [22, 40].

ENplus Certification System

Prior to the implementation of the ENplus Certification System in 2011, European, Canadian, and US pellet-producing companies had significant variation in official country quality standards and guidelines [41–43]. We found only a few publications about pellet certification (Figure 2), presumably because of the lack of guideline cohesiveness and only recent development of the ENplus system. The ENplus System allows for convenient and effective compliance with the European standard EN 14961-2 [44]. The purpose of this certification system is to establish a standardized quality regime for wood pellets for heating and combined heat and power (CHP) up to 1 MW output power in residential, commercial, and public buildings [44]. The System will create a "level playing field" for pellet producers and will boost consumers' confidence that they are receiving a quality product [45]. Under ENplus standards operational processes including production, logistics, and delivery are controlled and made transparent by defining the requirements for technical facilities, operational procedures, and documentation [44]. This transparency allows for quick and easy problem identification and solving, therefore minimizing downtime of production facilities. The German Pellet Institute (DEPI) developed the ENplus System and licensed it to the European Pellet Council (EPC), which is an organization within the European Biomass Association (AEBIOM) [44]. The specifications of the System include three classes of pellet quality: ENplus-A1, ENPlus-A2, and EN-B [44]. ENplus-A1 is used in residential boilers or stoves and is the premium class of pellets, producing the least amount of ash and meeting the highest standards [41]. ENplus-A2 pellets produce a higher amount of ash during combustion and are used in larger boiler systems [41]. The industrial grade of pellets under ENplus is classed as EN-B [41].

Table 1 summarizes the spectrum of the crucial pellet parameters for each class. Additives to improve fuel quality must not exceed 2% of the total mass of the pellets (≤1.8% of the total pellet mass in production and ≤0.2% of the total pellet mass after production) [44]. Each certified producer (and trader) must display the ENplus certification seal on their product [44]. Producers and traders of wood pellets that have adopted ENplus certification standards are found in countries around the world including Austria, Belgium, Canada, Croatia, Czech Republic, Denmark, France, Germany, Italy, Lithuania, Poland, Portugal, Romania, Slovenia, Slovakia, Spain, Switzerland, the Netherlands, the US, and the UK [41]. Canada is making the switch to the ENplus standard; The Wood Pellet Association of Canada has already applied to receive the ENplus license for Canada [46].

Table 1: Ranges of EN 14961-2 values for the most crucial wood pellet parameters

Property	Unit[*]	ENplus-A1	ENplus-A2	EN-B	Testing standard
Diameter	mm	6 or 8	6 or 8	6 or 8	EN 16127
Length	mm	$3.15 \leq L \leq 40$[3]	$3.15 \leq L \leq 40$[3]	$3.15 \leq L \leq 40$[3]	EN 16127
Moisture content	w-%[1]	≤10	≤10	≤10	EN 14774-1

Ash content	w-%[2]	≤0.7	≤1.5	≤3.0	EN 14775 (550°C)
Mechanical durability	w-%[1]	≥97.5[4]	≥97.5[4]	≥96.5 [4]	EN 15210-1
Fines (<3.15 mm)	w-%[1]	<1	<1	<1	EN 15210-1
Net calorific value	MJ/kg[1]	16.5 ≤ Q ≤ 19	16.3 ≤ Q ≤ 19	16.0 ≤ Q ≤ 19	EN 14918
Bulk density	kg/m³	≥600	≥600	≥600	EN 15103
Nitrogen content	w-%[2]	≤0.3	≤0.5	≤1.0	EN 15104
Sulfur content	w-%[2]	≤0.03	≤0.03	≤0.04	EN 15289
Chlorine content	w-%[2]	≤0.02	≤0.02	≤0.03	EN 15289
Ash melting behaviour[4]	°C	≥1200	≥1100	≥1100	EN 15370

As received; [2]dry basis; [3]a maximum of 1 w-% of the pellets may be longer than 40 mm; no pellets >45 mm are allowed; [4]deformation temperature; sample preparation at 815 C; [*]w-%: percentage of total pellet mass. Source: [44].

Objective (ii): Value Chain and Supply Chain Management

As demonstrated in Figure 3, many studies were found on supply chain and value chain management. However, very few studies specifically described the value chain [8, 18, 47]. The value chain of wood pellet manufacturers includes the determination of the value associated with each stage of the supply chain, which includes raw material procurement, inbound logistics of raw materials, processing of raw materials into pellets, and outbound logistics to the end consumer [24]. Procurement includes the location of raw materials, the species of raw materials in existence for wood pellet usage, and the original state of these accessible materials (i.e., roundwood, wood chips, sawdust, or wood shavings). Inbound logistics is the method of transportation required to move the raw material from its original location to the manufacturing plant for processing and includes scheduling decisions.

Processing includes drying, grinding, pressing, cooling, and bagging/ storage. Outbound logistics is the method of transportation used to deliver the pellets to the end consumer and also includes scheduling decisions.

Transportation scheduling (logistics) is a very important component of the wood pellet value chain, as fuel prices and operator wages continue to increase, therefore requiring optimal transportation decisions to be made to avoid unnecessary costs. Pettersson and Segerstedt [48] define logistics cost as "cost components related to distribution or transportation cost and cost for warehouses," in an attempt to offer clarity to the term and separate it from the term "supply chain cost," which they define as "all relevant costs in the supply chain of the company or organisation in question." In an expansive nation such as Canada, it is not feasible to transport cutter shavings, sawdust, and/or wood chips over long distances [4, 6]. It is worthwhile to transport the densified wood pellets, as they have a high BTU/ volume ratio; however, the longer the haul distance for raw materials or finished pellets is, the less cost-effective it is for the producer [6]. Well-developed, seamless connections to marketing-sales and order- delivery processes are needed for efficient, cost-effective value chain coordination (i.e., backhauling) [49–51]. Rail transport is a very efficient and cost-effective means of moving wood pellets; however, not all producers have direct access to railways. Rail [52] has begun to more aggressively market their wood pellet transportation opportunities to Canadian producers. They offer the flexibility to ship wood pellets in bulk, bags, boxcars, and intermodal containers. Rail [52] ships over 800,000 t of wood pellets annually and is ranked as "North America's number 1 mover of forest products."

Uncertainty, Leanness, and Agility

When considering modeling of the (wood pellet) value chain within the manufacturing firm, agility must be achieved to account for differences in specifications and types of wood pellets, as well as differences in procurement, processing, and distribution methods and locations. Value chain models should be created with the intent to readily change these inputs based on market demand and should reduce operational planning cycles [49]. Operational planning cycles include all activities that must be planned to ensure successful operation of a business in a very short-term time period (i.e., one week) [49]. In recent years the trend in supply chain management has been to make supply chains (and their integrated value chains) more agile, flexible, and responsive [53–57]. This trend is illustrated by Figure 3, which shows that we reviewed 14 publications specifically addressing these supply chain attributes.

Mathematical models have been used as decision support tools to assist managers in decision-making processes for strategic, tactical, and operational level planning. Operational-level management must focus on short-term productivity and process optimization to meet changing market trends [58]. Cost forecasting under uncertainty can lead to inaccurate model results; therefore, uncertainty must be remedied by increasing clarity and accuracy of input information [59]. "Decisions are made under certainty when perfect information is available and under uncertainty when one has only partial (or imperfect) information" [50]. Deterministic models serve as a "solid foundation" for value chain network design; however, deterministic models do not handle uncertainties and changes in information [50]. Therefore, stochastic models must be used which take into account realistic factors that affect business operations, including (but not limited to) raw material prices, energy costs, market demand for the end product(s), the cost of labour, retail price(s) of the finished product(s), and exchange rates [50, 60].

Leanness in a firm's value chain refers to its ability to "do more with less" and minimize (or eliminate) waste in its operations with cost leadership and cost performance strategies [61–64]. Agility in a firm's value chain encompasses operational flexibility performance and responsiveness to changes in information, such as product volume and/or logistics scheduling fluctuations [65–68]. Agility must also be applied not only within the individual firm's value chain but also throughout the supply chain as part of the partner selection process to create agile supply chains [69]. Both agility and leanness are strategies useful for developing or maintaining a competitive advantage in an uncertain marketplace.

Managerial Involvement

During this review, we uncovered six papers addressing the importance of managerial involvement in value chain and supply chain optimization [70–72] (Figure 3). Supply and demand are dynamic and ongoing processes; and, therefore, managing the value chain of a company should be considered as an on-going relationship between suppliers, the manufacturing firm and, end consumers [71]. The more involved management becomes with the value chain, the more they may visualize linkages of the value chain with the overall marketing strategy and goals of the firm, and the more likely management is to follow through with the successful application of the value chain at the operational level [72]. Gooch [70] found that, even when value chain optimization strategies are implemented within a firm, human resistance is inevitable and can seriously detract from the effectiveness of the management plan. Sometimes managers are even confused by the propositions and/or realize that the value chain model is being incorrectly used and therefore refuse to support it rather than work to improve its usage [64]. Unfortunately this resistance to change seriously affects the performance of the firm.

When dealing with complex value chains, identification of the critical value network locations is a useful managerial approach [73]. Lind et al. [13] emphasized that managing the working capital (short-term finance flow) of a company and its supply chain should be a major focus instead of just managing the flow of goods through the supply chain. Cantor and Macdonald [74] reviewed management problem-solving approaches within the supply chain and found that a more abstract approach to decision making may actually achieve better overall results than a more concrete approach. Cantor and Macdonald [74] discussed the fact that having complex, system-wide knowledge (more information) overwhelmed many managers, leading to poor decision making. Therefore, the use of a decision support system can simplify the overload of information and help managers make better decisions while still having all available information at their fingertips.

Demand-driven Management Approach

Demand-driven supply networks aim to link the supply/production rate directly to the level of actual demand for a specified time period in order to enable the manufacturer to respond in real-time to shifts in the level of demand and gain insight into general demand trends for their product(s) [49, 75]. The "upstream" component of the manufacturing value chain is the origin of the raw materials used to produce a product and the transportation of these materials to the processing facility, while the "downstream" component of the value chain follows processing to distribution of the final product to the end consumers [76]. Most companies, by default, examine their supply chains and value chains from an "upstream to downstream" perspective (as a directional flow), meaning that they operate by creating the product based on capacity, with some concept of forecasted market demand, and "push" the product out into the marketplace and also examine associated value creation in this manner [77–79]. Neumann et al. [62] conducted surveys of various companies only to find that even when it came to incorporating lean production techniques, very few companies used a demand-pull strategy. The findings of this review are in accordance with those of Neumann et al. [62], as only six (9%) of the 68 papers studied for objective 2 discussed the effectiveness of demand-driven management approaches [49, 75, 76, 80–82] (Figure 3). Demand Forecasting approaches were also considered by three of the six papers in objective 2 addressing demand-driven management [49, 75,82].

Demand-driven management adopts the value chain's downstream to upstream perspective (as a directional flow) and applies it to the supply chain. This application allows for production to become a reactive process based on the signals sent by real-time demand to the upstream (procurement) end of the supply chain and the product is "pulled" through the supply chain and/or value chain by the quantity demanded, instead of being "pushed" out into the market [83, 84]. Demand-driven models, used to support demand-driven management techniques, are very advantageous for many reasons. These reduce, or eliminate, inefficiencies throughout the supply chain and allow for a "smooth product flow", have shown significant improvements in utilization processes, improve inventory management and achieve optimal production capacities, and are more successful at responding to supply fluctuations [84, 85].

Demand Forecasting

Demand forecasts are crucial to provide input for demand-driven planning systems. Multiple approaches are available to forecast demand. Vinterbäck [86], and Hosoda, and Disney [87] discuss some of these approaches, including exponential smoothing, the naïve approach, moving average, autoregressive (AR), autoregressive integrated moving average (ARIMA), autoregressive extra (ARX), vector autoregressive (VAR), neural networks and the quantile regression method. However, these methods have not been proven to be overly effective and still allow for inaccurate demand prediction at each level throughout the supply chain, resulting in the bullwhip effect, which is amplification in demand variability when moving upstream through a value chain or supply chain [88]. Therefore, we are considering the new approaches currently being researched to increase the forecasting accuracy. Multilayer perception (MLP) is an approach that generalizes either linear or nonlinear functional relationships between inputs and outputs [88]. Yousefi et al. [82] designed a comprehensive demand response (CDR) model for a Retail Energy Provider agent in an agent-based retail environment to offer real-time energy prices to customers. Yousefi et al. [82] found that the CDR model gave a better representation of customers' historical behavior for future demand prediction. We only found four papers in our review that addressed demand forecasting [50, 79, 87, 88], while another three papers, which we categorized as demand-driven management papers, also addressed demand forecasting approaches [49, 75, 82].

Demand for Wood Pellets and Other Biofuels

Of the 68 papers reviewed for objective 2, 33 (49%) considered market demand for wood pellets and other biofuels (Figure 3). North America began producing wood pellets for a small niche market in the 1930s, with a significant market growth spurt occurring in the 1970s, followed by rapid market development in the 1990s as a result of increased consumption in Europe [89, 90]. In Canada in 1997 pellet production was only 173,000 t, of which roughly two-thirds were exported to the US, but from 1997 to 2007 Canada went from exporting 0% to 63% of its pellets to the European market, which displaced the US from its position as Canada's major trade partner [3]. In 2010, wood pellet production was less than 70% of design capacity in Europe, implying a lack of natural resources for pellet production, and, therefore, indicating a need for pellet import [5]. Imported biomass comprises between 21% and 43% of Europe's total available biomass [6]. Canada is now one of the world's leaders with regard to production and trade success of wood pellets because of many contributing factors, including its surplus of natural resources, low-cost mill residue, excess pellet production capacity, and abundance of export opportunities [91–94].

Obernberger and Thek's [3] prognosis for Canada was for 5.5 Mt to be produced in 2010. However, the production capacity of Canada in 2010 was only 2.08 Mt per year and in 2011 it expanded to 3.22 Mt per year (a 55% growth from 2010 to 2011). However, not all production plants are (or were) operating at full capacity due to market conditions [95, 96].

Market studies on Canada and other relevant countries show that Canada is lacking in domestic wood pellet demand as compared with these other countries; therefore most of Canada's pellet production is exported [95, 97]. However, some of these studies have noted that there is a rising trend in Canada's domestic consumption of wood pellets and that Canada has great potential for growth with regard to domestic pellet consumption [92, 94, and 98]. Junginger et al. [6, 98] identify logistics as the most influential trade barrier for wood pellets, while development of technical standards presents itself as a major opportunity for wood pellet trade. Wolf et al. [31] identified the need to more efficiently produce biomass in order to meet expanding market demand and studied the effectiveness and feasibility of biofuel production in the forestry industry.

There has been rapid growth in the worldwide production and consumption of wood pellets and other biomasses within the last decade [99, 100]. Canada, the US, Korea, and countries throughout Europe exhibit this global trend [101–110]. A factor contributing to the onset of this trend is favourable government policy implementation, which has allowed for an effective increase in pellet production and consumption [99]. Provincial governments throughout Canada have successfully implemented various initiatives to promote renewable energy production and usage. For example, Ontario's Green Energy Act of 2009 applied a Feed-in-Tariff (FIT) program that offers price incentives for new electrical generating stations that are fueled by renewable resources [111]. Aboriginal, or first nation, communities in Ontario have also begun the process of adopting renewable energy initiatives. The community of Pic River First Nation has various current and future renewable energy projects and is actively participating in knowledge and information sharing with other first nation's communities across Ontario and Canada [112].

Average worldwide demand (consumption) for wood pellets increased from 3.28 Mt in 2003 to 10.54 Mt in 2007 (a 41.7% increase), average worldwide production increased from 3.38 Mt in 2003 to 10.54 Mt in 2007 (a 40.5% increase), and average worldwide capacity increased from 4.5 Mt in 2003 to 15.0 Mt in 2007 (a 43.1% increase) [5]. Sweden is one of the world's largest producers and consumers of wood pellets, due mainly to its favourable taxation system towards biofuels, ubiquitous district heating systems, and abundance of raw materials [97]. Pellet usage for heating/energy allows for improved fuel supply security (from a renewable resource viewpoint) and stimulates local and regional job creation and overall economic development [110]. Generally, the availability of forest resources, the demand for forest fuels, and machine and labour costs are the defining factors behind (wood pellet) prices [32, 34]. Other factors contributing to the global success of the wood pellet industry include the automation of heating systems, logistics infrastructure, national funding systems coupled with marketing programs and public awareness campaigns, and price increases in the oil and gas sector [113]. As the marketplace expands and demand for wood pellets increases, if the demand for pellets exceeds the current capacity of production plants, they will have to increase capacity in order to satisfy demand and remain competitive [32].

Objective (iii): Supply Chain and Value Chain Models

Papageorgiou [60] identifies two broad categories for supply chain models: mathematical programming models and simulation models. He explains that mathematical models are used for optimizing high-level decisions with an aggregate view of operational processes, while simulation-based models are more accurate as they study detailed, dynamic operations under uncertainty.

Supply Chain Models

We discovered that many general supply chain models have been developed, covering a wide variety of products (Table 2). Biomass includes all plant and plant-derived materials (including animal manure!) that can be considered a part of the present carbon cycle [114]. Using this broad definition, there have also been many supply chain models created relating to biomass (Table 2).

Table 2: Categorization of model types in reviewed papers

Type of study		Broad modeling category	Type of modeling approach	Authors using the modeling approach
Supply chain models	Biomass supply chain models	Simulation	Simulation	[80, 128]
			Demand driven	[83, 84]
			Simulation-based fuzzy inventory	[129]
			Integrated biomass supply and logistics (ISBAL) modeling environment-dynamic simulation	[130]

Supply chain models	Biomass supply chain models	Mathematical programming	Dynamic, nonlinear mixed integer	[131, 132]
			Dynamic, linear mixed integer	[114, 133]
			Scenario-based optimization	[32, 125]
			Agent-based models (ABMs)	[134]
			Land-suitability model (LSM) using analytic hierarchy process (AHP)	[135]
			Power function utilization	[33]
			Spatial partial equilibrium	[136]
			Technoeconomic	[36, 126]
			Game theoretic approach	[127]
			Mixed integer	[137]
			Mixed integer linear programming (MILP)	[138]
			Linear programming network	[139]
			Process network synthesis (PNS) two-level process graph (P-graph) approach	[31, 140]
			Newsvendor economic	[141]
			Integrated optimization	[142]
Supply chain models	General supply chain models	Simulation	Simulation	[21, 143, 144]

Supply chain models	General supply chain models	Mathematical programming	Demand driven	[83]
			Agent-based models (ABMs)	[78]
			Closed-loop optimization	[53]
			Mixed integer	[145]
			Lead-time inventory	[65, 146, 147]
			Stochastic network	[148]
			Fuzzy programming	[149]
			Mixed integer linear programming (MILP)	[150–152]
			Integrated business planning (IBP) matrix	[153]
			Genetic algorithms	[57, 154, 155]
Value chain models	General value chain models	Mathematical programming	Mixed integer nonlinear programming	[115, 156]
			Object oriented programming approach with ecological mass balance	[116]
			Analytic network process (ANP)	[47]
Value chain models	Wood pellet value chain models	Simulation	Inventory management (Pell-Sim)	[86]
Value chain models	Wood pellet value chain models	Mathematical programming	Win-win optimization	[113]
			Technoeconomic	[30]
			Static partial equilibrium	[24]
Other	Wood pellet theses	Mathematical programming	Linear multicommodity network flow	[123]
			Scenario-based financial	[122, 124]
Other	Biomass business plans	Mathematical programming	Financial-based	[117, 118]

Other	Wood pellet business plans	Mathematical programming	Financial-based	[119, 120]

Value Chain Models

We found that relatively few academic research papers have been published specifically on value chain modeling [115, 116]. There were three studies found specifically on wood pellet value chain analysis [24, 30,113]. Other sources found were five feasibility studies for actual biomass and wood pellet production facilities [117–121] and three university theses on wood pellet production and feasibility [122–124]. Refer to Table 2 below for a comprehensive list.

Demand-driven Models

Our review uncovered only two papers discussing and utilizing demand-driven approaches to modeling [83,84] and these were both supply chain models (Table 2).

Objective (iv): Perspective

This section explains the gaps in the existing literature based on the findings of our review for each objective. We identify suggestions for future studies that may successfully fill in these gaps.

Literature Gaps and Recommendations for Future Research

There is a need for more (Canadian) studies specifically about wood pellet production methods and characteristics. The low number of Canadian-based value chain and supply chain studies indicates a rather large gap that needs to be filled as well. There is a need for more value chain models in general but especially those relating to wood pellet production. Going hand-in-hand with the value chain gap is the gap relating to managerial involvement, as defined in Section 3.2.2.

Future studies focused on value chain optimization can be paired with guidelines for convenient and effective managerial execution.

There is also a great need for dynamic, demand-driven models within the value chain and the supply chain. This need may be coupled with the necessity of more accurate and effective demand forecasting methods. By employing capacity-optimization techniques similar to those outlined in previous studies [32, 125–127] pellet production costs may be minimized as a function of plant capacity, utilizing real-time information and emulating stochastic market demand. Following the lead of Trapero et al. [88] and Yousefi et al. [82] and building upon their results would produce cutting-edge demand forecasting methods to improve demand-driven modeling approaches.

CONCLUSIONS

The available literature associated with wood pellet production and characteristics, value/supply chain management concerns and value/supply chain modeling techniques was explored in this review paper. The results show that relatively few published studies provide an in-depth account of wood pellet production, pellet plant feasibility and wood pellet value/supply chain management. There are also very few papers explaining the overall concept of the value chain. We discovered only a small number of Canadian-based studies during this review as well. Future studies to fill these gaps, particularly within the realm of the value chain for Canadian pellet-producing companies, should be conducted. These studies would not only be beneficial to the producers but also to the companies linked to the producers within the supply chain. Currently, measures are being taken to expand Canada's bioeconomy; therefore new studies demonstrating feasible projects relating to the bioeconomy, with improved supply chains and value chains, are essential to promote this expansion. Implementation of modern demand forecasting and demand-driven supply chain management techniques will improve the operations of current facilities and most effectively manage the operations of new facilities.

REFERENCES

1.   E. Alakangas and M. Virkkunen, EUBIONET II—Biomass Fuel Supply Chains For Solid biofuels, From Small To Large Scale, VTT Technical Research Centre of Finland, Jyväskylä, Finland, 2007.

2.   M. Kennedy, R. Wong, A. Vandenbroek, D. Lovekin, and M. Raynolds, Biomass Sustainability Analysis, An Assessment of Ontario-Sourced Forest-Based Biomass For Electricity Generation, The Pembina Institute, Alberta, Canada, 2011.

3.   I. Obernberger and G. Thek, The Pellet Handbook, the Production and thermal Utilisation of Pellets, Earthscan, London, UK, 2010.

4.   W. Rickerson, T. Halfpenny, and S. Cohan, "The emergence of renewable heating and cooling policy in the United States," Policy and Society, vol. 27, no. 4, pp. 365–377, 2009.

5.   J. H. Peng, H. T. Bi, S. Sokhansanj, J. C. Lim, and S. Melin, "An economical and market analysis of Canadian wood pellets," International Journal of Green Energy, vol. 7, no. 2, pp. 128–142, 2010.

6.   M. Junginger, T. Bolkesjø, D. Bradley et al., "Developments in international bioenergy trade," Biomass and Bioenergy, vol. 32, no. 8, pp. 717–729, 2008.

7.   E. K. Ackom, W. E. Mabee, and J. N. Saddler, "Industrial sustainability of competing wood energy options in Canada," Applied Biochemistry and Biotechnology, vol. 162, no. 8, pp. 2259–2272, 2010.

8.   M. C. Mahutga, "When do value chains go global? A theory of the spatialization of global value chains," Global Networks, vol. 12, no. 1, pp. 1–21, 2012.

9.   M. Porter, Competitive Advantage: Creating and Sustaining Superior Performance, Collier-Macmillan, New York, NY, USA, 1985.

10.   D. W. te Velde, J. Rushton, K. Schreckenberg et al., "Entrepreneurship in value chains of non-timber forest products," Forest Policy and Economics, vol. 8, no. 7, pp. 725–741, 2006.

11.   D. Walters, "Competition, collaboration, and creating value in the value chain," in Modelling Value, Contributions To Management Science: Selected Papers of the 1st International Conference on

Value Chain Management, H. Jodlbauer, J. Olhager, and R. J. Schonberger, Eds., pp. 3–36, University of Applied Sciences in Upper Austria, School of Management, Steyr, Austria, 2012.

12. A. K. N. Aoudji, A. Adégbidi, V. Agbo et al., "Functioning of farm-grown timber value chains: lessons from the smallholder-produced teak (Tectona grandis L.f.) poles value chain in Southern Benin,"Forest Policy and Economics, vol. 15, pp. 98–107, 2012.

13. L. Lind, M. Pirttilä, S. Viskari, F. Schupp, and T. Kärri, "Working capital management in the automotive industry: financial value chain analysis," Journal of Purchasing and Supply Management, vol. 18, no. 2, pp. 92–100, 2012.

14. A. Booker, D. Johnston, and M. Heinrich, "Value chains of herbal medicines—research needs and key challenges in the context of ethnopharmacology," Journal of Ethnopharmacology, vol. 140, no. 3, pp. 624–633, 2012.

15. R. Sathre and L. Gustavsson, "Process-based analysis of added value in forest product industries,"Forest Policy and Economics, vol. 11, no. 1, pp. 65–75, 2009.

16. J. T. Mentzer, W. DeWitt, J. S. Keebler et al., "Defining supply chain management," Journal of Business Logistics, vol. 22, no. 2, pp. 1–25, 2001.

17. J. von Geibler, K. Kristof, and K. Bienge, "Sustainability assessment of entire forest value chains: integrating stakeholder perspectives and indicators in decision support tools," Ecological Modelling, vol. 221, no. 18, pp. 2206–2214, 2010.

18. D. Rana and M. Gregory, "Exploring the role of business support agencies in value chain management of the medical device industry," in Modelling Value, Contributions To Management Science: Selected Papers of the 1st International Conference on Value Chain Management, H. Jodlbauer, J. Olhager, and R. J. Schonberger, Eds., pp. 393–416, University of Applied Sciences in Upper Austria, School of Management, Steyr, Austria, 2012.

19. K. Arthofer, C. Engelhardt-Nowitzki, H. P. Feichtenschlager, and D. Girardi, "Servicing individual product variants within value chains with an ontology," in Modelling Value, Contributions To Management Science: Selected Papers of the 1st International Conference on Value Chain Management, H. Jodlbauer, J. Olhager, and R. J. Schonberger, Eds., pp. 331–352, University

of Applied Sciences in Upper Austria, School of Management, Steyr, Austria, 2012.

20. G. Macfadyen, A. M. Nasr-Alla, D. Al-Kenawy et al., "Value-chain analysis—an assessment methodology to estimate Egyptian aquaculture sector performance," Aquaculture, vol. 362-363, pp. 18–27, 2012.

21. J. Venkateswaran and Y. J. Son, "Impact of modelling approximations in supply chain analysis—an experimental study," International Journal of Production Research, vol. 42, no. 15, pp. 2971–2992, 2004.

22. M. T. Hansen, A. R. Jein, S. Hayes, and P. Bateman, English Handbook For Wood Pellet Combustion, Pelletsatlas, Denmark, 2009.

23. S. Mani, S. Sokhansanj, X. Bi, and A. Turhollow, "Economics of producing fuel pellets from biomass,"Applied Engineering in Agriculture, vol. 22, no. 3, pp. 421–426, 2006

24. M. Mäkelä, J. Lintunen, H.-L. Kangas, and J. Uusivuori, "Pellet promotion in the Finnish sawmilling industry: the cost-effectiveness of different policy instruments," Journal of Forest Economics, vol. 17, no. 2, pp. 185–196, 2011.

25. N. Saracoglu and G. Gunduz, "Wood pellets—tomorrow's fuel for Europe," Energy Sources A, vol. 31, no. 19, pp. 1708–1718, 2009.

26. H. Spelter and D. Toth, North America's Wood Pellet Sector, Department of Agriculture, Forest Service, Forest Products Laboratory, Madison, Wis, USA, 2009.

27. A. Tapaninen, Adoption of Innovation: Wood Pellet Heating System in the Renewable Residential Energy Context, Faculty of Business and Technology Management. Tampere University of Technology, Tampere, Finland, 2010.

28. P. Åsman, Pellet Tool Kit, A Basic How-To Guide Prior To Starting Your Pellet Project, Northern Ontario Value Added Initiative, FP Innovations, Timmins, Ontario.

29. A. Uasuf and G. Becker, "Wood pellets production costs and energy consumption under different framework conditions in Northeast Argentina," Biomass and Bioenergy, vol. 35, no. 3, pp. 1357–1366, 2011.

30. A. Pirraglia, R. Gonzalez, and D. Saloni, "Techno-economical analysis of wood pellets production for U.S. manufacturers," BioResources, vol. 5, no. 4, pp. 2374–2390, 2010.

31. A. Wolf, A. Vidlund, and E. Andersson, "Energy-efficient pellet production in the forest industry—a study of obstacles and success factors," Biomass and Bioenergy, vol. 30, no. 1, pp. 38–45, 2006.

32. D. Alfonso, C. Perpiñá, A. Pérez-Navarro, E. Peñalvo, C. Vargas, and R. Cárdenas, "Methodology for optimization of distributed biomass resources evaluation, management and final energy use," Biomass and Bioenergy, vol. 33, no. 8, pp. 1070–1079, 2009.

33. P. W. Gallagher, H. Brubaker, and H. Shapouri, "Plant size: capital cost relationships in the dry mill ethanol industry," Biomass and Bioenergy, vol. 28, no. 6, pp. 565–571, 2005.

34. K. Mahapatra, L. Gustavsson, and R. Madlener, "Bioenergy innovations: the case of wood pellet systems in Sweden," Technology Analysis and Strategic Management, vol. 19, no. 1, pp. 99–125, 2007

35. N. Kaliyan and R. Vance Morey, "Factors affecting strength and durability of densified biomass products," Biomass and Bioenergy, vol. 33, no. 3, pp. 337–359, 2009.

36. A. Sultana, A. Kumar, and D. Harfield, "Development of agri-pellet production cost and optimum size," Bioresource Technology, vol. 101, no. 14, pp. 5609–5621, 2010.

37. A. Pa, X. T. Bi, and S. Sokhansanj, "A life cycle evaluation of wood pellet gasification for district heating in British Columbia," Bioresource Technology, vol. 102, no. 10, pp. 6167–6177, 2011.

38. A. Sultana and A. Kumar, "Ranking of biomass pellets by integration of economic, environmental and technical factors," Biomass and Bioenergy, vol. 39, pp. 344–355, 2012.

39. S. Naik, V. V. Goud, P. K. Rout, K. Jacobson, and A. K. Dalai, "Characterization of Canadian biomass for alternative renewable biofuel," Renewable Energy, vol. 35, no. 8, pp. 1624–1631, 2010.

40. I. Obernberger and G. Thek, "Physical characterisation and chemical composition of densified biomass fuels with regard to their combustion behaviour," Biomass and Bioenergy, vol. 27, no. 6, pp. 653–669, 2004.

41. AEBIOM (European Biomass Association), 2013, ENplus, AEBIOM, Brussels, Belgium,http://www.enplus-pellets.eu/.

42. A. García-Maraver, V. Popov, and M. Zamorano, "A review of European standards for pellet quality,"Renewable Energy, vol. 36, no. 12, pp. 3537–3540, 2011.

43. PFI (Pellet Fuels Institute), Pellet Fuels Institute Standard Specification For Residential/Commercial Densified Fuel, PFI, Arlington, Va, USA, 2011.

44. EPC (European Pellet Council), EN Plus: European Pellet Council Handbook For the Certification of Wood Pellets For Heating Purposes, Version 2.0., European PelletCouncil, European Biomass Association, Brussels, Belgium, 2013.

45. WPAC (Wood Pellet Association of Canada), Sustainable Biomass Production, Annex Business Media, 2013, http://www.pellet.org/wpac-news/sustainable-biomass-production.

46. AEBIOM (European Biomass Association), ENplus Newsletter, AEBIOM, Brussels, Belgium, 2012.

47. G. Kayakutlu and G. Buyukozkan, "Assessing performance factors for a 3PL in a value chain,"International Journal of Production Economics, vol. 131, no. 2, pp. 441–452, 2011.

48. A. I. Pettersson and A. Segerstedt, "Measuring supply chain cost," International Journal of Production Economics, vol. 143, no. 2, pp. 357–363, 2012.

49. M. Panley and S. Boerner, Demand-Driven Supply Networks: Advancing Supply Chain Management, SAP AG, Germany, 2006.

50. W. Klibi, A. Martel, and A. Guitouni, "The design of robust value-creating supply chain networks: a critical review," European Journal of Operational Research, vol. 203, no. 2, pp. 283–293, 2010.

51. S. Gold and S. Seuring, "Supply chain and logistics issues of bio-energy production," Journal of Cleaner Production, vol. 19, no. 1, pp. 32–42, 2011.

52. C. N. Rail, 2012, Wood Pellets, http://www.cn.ca/en/shipping-alternative-energy-products-wood-pellets.htm.

53. M. S. Pishvaee, M. Rabbani, and S. A. Torabi, "A robust optimization approach to closed-loop supply chain network

design under uncertainty," Applied Mathematical Modelling, vol. 35, no. 2, pp. 637–649, 2011.

54. J. Godsell, T. Diefenbach, C. Clemmow, D. Towill, and M. Christopher, "Enabling supply Chain segmentation through demand profiling," International Journal of Physical Distribution and Logistics Management, vol. 41, no. 3, pp. 296–314, 2011.

55. D. Ivanov, B. Sokolov, and J. Kaeschel, "A multi-structural framework for adaptive supply chain planning and operations control with structure dynamics considerations," European Journal of Operational Research, vol. 200, no. 2, pp. 409–420, 2010.

56. F. Pan and R. Nagi, "Robust supply chain design under uncertain demand in agile manufacturing,"Computers and Operations Research, vol. 37, no. 4, pp. 668–683, 2010.

57. A. D. Yimer and K. Demirli, "A genetic approach to two-phase optimization of dynamic supply chain scheduling," Computers and Industrial Engineering, vol. 58, no. 3, pp. 411–422, 2010.

58. A. Gunasekaran and E. W. T. Ngai, "The future of operations management: an outlook and analysis,"International Journal of Production Economics, vol. 135, no. 2, pp. 687–701, 2012. View at Publisher ·View at Google Scholar · View at Scopus

59. M. E. Kreye, Y. M. Goh, L. B. Newnes, and P. Goodwin, "Approaches to displaying information to assist decisions under uncertainty," Omega, vol. 40, no. 6, pp. 682–692, 2012.

60. L. G. Papageorgiou, "Supply chain optimisation for the process industries: advances and opportunities," Computers and Chemical Engineering, vol. 33, no. 12, pp. 1931–1938, 2009.

61. M. Christopher and D. R. Towill, "Supply chain migration from lean and functional to agile and customised," Supply Chain Management, vol. 5, no. 4, pp. 206–213, 2000.

62. C. Neumann, S. Kohlhuber, and S. Hanusch, "Lean production Austrian industrial companies: an empirical investigation," in Modelling value, Contributions to Management Science: selected papers of the 1st International Conference on Value Chain Management, H. Jodlbauer, J. Olhager, and R. J. Schonberger, Eds., pp. 293–312, University of Applied Sciences in Upper Austria, School of Management, Steyr, Austria, May 2012.

63. J. Olhager and D. I. Prajogo, "The impact of manufacturing and supply chain improvement initiatives: a survey comparing make-to-order and make-to-stock firms," Omega, vol. 40, no. 2, pp. 159–165, 2012.

64. R. J. Schonberger, "Measurement of lean value chains: efficiency and effectiveness," in Modelling value, Contributions to Management Science: selected papers of the 1st International Conference on Value Chain Management, H. Jodlbauer, J. Olhager, and R. J. Schonberger, Eds., pp. 65–75, University of Applied Sciences in Upper Austria, School of Management, Steyr, Austria, May 2012.

65. J. Blackburn, "Valuing time in supply chains: establishing limits of time-based competition," Journal of Operations Management, vol. 30, no. 5, pp. 396–405, 2012.

66. E. W. T. Ngai, D. C. K. Chau, and T. L. A. Chan, "Information technology, operational, and management competencies for supply chain agility: findings from case studies," Journal of Strategic Information Systems, vol. 20, no. 3, pp. 232–249, 2011.

67. J. M. Rudd, G. E. Greenley, A. T. Beatson, and I. N. Lings, "Strategic planning and performance: extending the debate," Journal of Business Research, vol. 61, no. 2, pp. 99–108, 2008.

68. P. Schütz and A. Tomasgard, "The impact of flexibility on operational supply chain planning," International Journal of Production Economics, vol. 134, no. 2, pp. 300–311, 2011.

69. C. Wu and D. Barnes, "A literature review of decision-making models and approaches for partner selection in agile supply chains," Journal of Purchasing and Supply Management, vol. 17, no. 4, pp. 256–274, 2011.

70. M. Gooch, "Evaluating the effectiveness of experiential learning for motivating value chain stakeholders to adopt new ways of capturing value," in Proceedings of the 1st International Conference on Value Chain Management Modelling value, Contributions to Management Science, H. Jodlbauer, J. Olhager, and R. J. Schonberger, Eds., pp. 49–64, University of Applied Sciences in Upper Austria, School of Management, Steyr, Austria, May 2012.

71. J. Kraigher-Krainer, "Habit, affect and cognition: a constructivist model on how they shape decision making," in Proceedings of

the 1st International Conference on Value Chain Management Modelling value, Contributions to Management Science, H. Jodlbauer, J. Olhager, and R. J. Schonberger, Eds., pp. 189–206, University of Applied Sciences in Upper Austria, School of Management, Steyr, Austria, May 2012.

72. C. Öberg, "What happened with the grandiose plans? Strategic plans and network realities in B2B interaction," Industrial Marketing Management, vol. 39, no. 6, pp. 963–974, 2010.

73. C. Engelhardt-Nowitzki, K. Arthofer, N. Kryvinska, and C. Strauss, "Supporting value chain integration through ontology-based modeling," in Proceedings of the 6th International Conference on Complex, Intelligent, and Software Intensive Systems, L. Barolli, F. Xhafa, S. Vitabile, and M. Uehara, Eds., IEEE Computer Society, Sanpaolo Palace Hotel, Palermo, Italy, July 2012.

74. D. E. Cantor and J. R. Macdonald, "Decision-making in the supply chain: examining problem solving approaches and information availability," Journal of Operations Management, vol. 27, no. 3, pp. 220–232, 2009.

75. Subramanian, B. G. P. S, and V. S. Reddy, "Transforming data driven SCM to demand driven SCM through lead time optimization," in Ninth AIMS International Conference on Management, Pune, India, January 2012.

76. H. An, W. E. Wilhelm, and S. W. Searcy, "Biofuel and petroleum-based fuel supply chain research: a literature review," Biomass and Bioenergy, vol. 35, no. 9, pp. 3763–3774, 2011.

77. C. Chandra and S. Kumar, "Supply chain management in theory and practice: a passing fad or a fundamental change?" Industrial Management and Data Systems, vol. 100, no. 3, pp. 100–113, 2000.

78. K. J. Mizgier, S. M. Wagner, and J. A. Holyst, "Modeling defaults of companies in multi-stage supply chain networks," International Journal of Production Economics, vol. 135, no. 1, pp. 14–23, 2012.

79. A. Toppinen and J. Kuuluvainen, "Forest sector modelling in Europe-the state of the art and future research directions," Forest Policy and Economics, vol. 12, no. 1, pp. 2–8, 2010.

80. M. Singer and P. Donoso, "Upstream or downstream in the value chain?" Journal of Business Research, vol. 61, no. 6, pp. 669–677, 2008.

81. C. Li, "Toward full, multiple, and optimal wood fibre utilization: a modeling perspective," Forestry Chronicle, vol. 85, no. 3, pp. 377–381, 2009.

82. S. Yousefi, M. P. Moghaddam, and V. J. Majd, "Optimal real time pricing in an agent-based retail market using a comprehensive demand response model," Energy, vol. 36, no. 9, pp. 5716–5727, 2011

83. N. Ayoub and N. Yuji, "Demand-driven optimization approach for biomass utilization networks,"Computers and Chemical Engineering, vol. 36, no. 1, pp. 129–139, 2012

84. T. Wöhrle, "Supply chain at western digital," Supply Chain Europe, vol. 18, no. 3, pp. 46–47, 2009.

85. N. Ayoub, H. Seki, and Y. Naka, "Superstructure-based design and operation for biomass utilization networks," Computers and Chemical Engineering, vol. 33, no. 10, pp. 1770–1780, 2009.

86. J. Vinterbäck, "Pell-Sim—dynamic model for forecasting storage and distribution of wood pellets,"Biomass and Bioenergy, vol. 27, no. 6, pp. 629–643, 2004.

87. T. Hosoda and S. M. Disney, "A delayed demand supply chain: incentives for upstream players,"Omega, vol. 40, no. 4, pp. 478–487, 2012.

88. J. R. Trapero, N. Kourentzes, and R. Fildes, "Impact of information exchange on supplier forecasting performance," Omega, vol. 40, no. 6, pp. 738–747, 2012.

89. B. Hillring and J. Vinterbäck, "Wood pellets in the Swedish residential market," Forest Products Journal, vol. 48, no. 5, pp. 67–72, 1998.

90. R. E. Löfstedt, "The use of biomass energy in a regional context: the case of Vaxjo Energi, Sweden,"Biomass and Bioenergy, vol. 11, no. 1, pp. 33–42, 1996.

91. E. Alakangas, M. Junginger, J. van Dam et al., "EUBIONET III-Solutions to biomass trade and market barriers," Renewable and Sustainable Energy Reviews, vol. 16, no. 6, pp. 4277–4290, 2012.

92. M. Cocchi, L. Nikolaisen, M. Junginger et al., Global Wood Pellet Industry Market and Trade Study, IEA Bioenergy, Task 40: Sustainable International Bioenergy Trade, Utrecht University, 2011.

93. L. Schroeder, Go Pellets Canada Pushes Domestic Market Sales, Canadian Bioenergy Association, Ottawa, Canada, 2011.

94. C. Verhoest and Y. Ryckmans, Industrial Wood Pellets Report, PellCert, 2012.

95. D. Bradley and K. Bradburn, Economic Impact of Bioenergy in Canada, Canadian Bioenergy Association, 2012.

96. D. Bradley and E. Thiffault, IEA Bioenergy Task 40 Country Report, Canada, Canadian Bioenergy Association, 2012.

97. M. Selkimäki, B. Mola-Yudego, D. Röser, R. Prinz, and L. Sikanen, "Present and future trends in pellet markets, raw materials, and supply logistics in Sweden and Finland," Renewable and Sustainable Energy Reviews, vol. 14, no. 9, pp. 3068–3075, 2010.

98. M. Junginger, J. van Dam, S. Zarrilli, F. Ali Mohamed, D. Marchal, and A. Faaij, "Opportunities and barriers for international bioenergy trade," Energy Policy, vol. 39, no. 4, pp. 2028–2042, 2011.

99. P. J. Ince, A. D. Kramp, K. E. Skog, D.-I. Yoo, and V. A. Sample, "Modeling future U.S. forest sector market and trade impacts of expansion in wood energy consumption," Journal of Forest Economics, vol. 17, no. 2, pp. 142–156, 2011.

100. P. Lamers, M. Junginger, C. Hamelinck, and A. Faaij, "Developments in international solid biofuel trade—an analysis of volumes, policies, and market factors," Renewable and Sustainable Energy Reviews, vol. 16, no. 5, pp. 3176–3199, 2012.

101. N. Lu and R. W. Rice, "Characteristics of wood fuel pellet manufacturers and markets in the united states, 2010," Forest Products Journal, vol. 61, no. 4, pp. 310–315, 2011.

102. V. Karkania, E. Fanara, and A. Zabaniotou, "Review of sustainable biomass pellets production—a study for agricultural residues pellets› market in Greece," Renewable and Sustainable Energy Reviews, vol. 16, no. 3, pp. 1426–1436, 2012.

103. E. Monteiro, V. Mantha, and A. Rouboa, "Portuguese pellets market: analysis of the production and utilization constrains," Energy Policy, vol. 42, pp. 129–135, 2012.

104. J.-H. Moon, J.-W. Lee, and U.-D. Lee, "Economic analysis of biomass power generation schemes under renewable energy initiative with Renewable Portfolio Standards (RPS) in Korea," Bioresource Technology, vol. 102, no. 20, pp. 9550–9557, 2011.

105. O. Olsson, B. Hillring, and J. Vinterbäck, "European wood pellet market integration—a study of the residential sector," Biomass and Bioenergy, vol. 35, no. 1, pp. 153–160, 2011.

106. B. M. Sopha, C. A. Klöckner, G. Skjevrak, and E. G. Hertwich, "Norwegian households› perception of wood pellet stove compared to air-to-air heat pump and electric heating," Energy Policy, vol. 38, no. 7, pp. 3744–3754, 2010.

107. E. Trømborg, M. Havskjold, O. Lislebø, and P. K. Rørstad, "Projecting demand and supply of forest biomass for heating in Norway," Energy Policy, vol. 39, no. 11, pp. 7049–7058, 2011.

108. J. van Dam, A. P. C. Faaij, J. Hilbert, H. Petruzzi, and W. C. Turkenburg, "Large-scale bioenergy production from soybeans and switchgrass in Argentina. Part A: potential and economic feasibility for national and international markets," Renewable and Sustainable Energy Reviews, vol. 13, no. 8, pp. 1710–1733, 2009.

109. V. K. Verma, S. Bram, and J. De Ruyck, "Small scale biomass heating systems: Standards, quality labelling and market driving factors—an EU outlook," Biomass and Bioenergy, vol. 33, no. 10, pp. 1393–1402, 2009.

110. J. Palladini, Canada›s Wood Products Industry, Conference Board of Canada, Ottawa, Canada, 2010.

111. A. Yatchew and A. Baziliauskas, "Ontario feed-in-tariff programs," Energy Policy, vol. 39, no. 7, pp. 3885–3893, 2011.

112. J. Krupa, "Blazing a new path forward: a case study on the renewable energy initiatives of the Pic River First Nation," Environmental Development, vol. 3, pp. 109–122, 2012.

113. V. Uran, "A model for establishing a win-win relationship between a wood pellets manufacturer and its customers," Biomass and Bioenergy, vol. 34, no. 5, pp. 747–753, 2010.

114. S. van Dyken, B. H. Bakken, and H. I. Skjelbred, "Linear mixed-integer models for biomass supply chains with transport, storage and processing," Energy, vol. 35, no. 3, pp. 1338–1350, 2010.

115. N. Shabani and T. Sowlati, "A mixed integer non-linear programming model for tactical value chain optimization of a wood biomass power plant," Applied Energy, vol. 104, pp. 353–361, 2013.

116. V. Christensen, J. Steenbeek, and P. Failler, "A combined ecosystem and value chain modeling approach for evaluating societal cost and benefit of fishing," Ecological Modelling, vol. 222, no. 3, pp. 857–864, 2011.

117. A. Oo, J. Kelly, and C. Lalonde, Assessment of Business Case For Purpose-Grown Biomass in Ontario, The Western University Research Park, Sarnia, Canada, 2012.

118. K. Campbell, A Feasibility Study Guide For An Agricultural Biomass Pellet Company, Agricultural Utilization Research Institute, St. Paul, Minn, USA, 2007.

119. BW McCloy & Associates, NWT Wood Pellet Pre-Feasibility Analysis, FPInnovations, Forintek Division, 2009.

120. G. Murray, Lillooet Biomass Energy Corporation Business Plan for a Wood Pellet Plant, Gordon Murray Corporate Finance, Revelstoke, Canada, 2010.

121. NEOS Corporation, Wood Pelletization Sourcebook: A Sample Business Plan For the Potential Pellet Manufacturer, Report No. DE-FG05-83R21390 (44), Great Lakes Regional Biomass Energy Program, 1995.

122. G. Blom, The feasibility of a wood pellet plant using alternate sources of wood fibre [B.Sc.F. thesis], University of British Columbia, Vancouver, Canada, 2009.

123. A. Ravn and S. P. Engstrøm, Modelling of wood pellet production and distribution for energy consumption [M.S. thesis], Technical University of Denmark, DTU Management Engineering, Kongens Lyngby, Denmark, 2010.

124. E. Urbanowski, Strategic analysis of a pellet fuel opportunity in Northwest British Columbia [M.B.A. thesis], Simon Fraser University,, 2005.

125. A. Kumar, J. B. Cameron, and P. C. Flynn, "Biomass power cost and optimum plant size in western Canada," Biomass and Bioenergy, vol. 24, no. 6, pp. 445–464, 2003.

126. B. M. Jenkins, "A comment on the optimal sizing of a biomass utilization facility under constant and variable cost scaling," Biomass and Bioenergy, vol. 13, no. 1-2, pp. 1–9, 1997.

127. F. Nasiri and G. Zaccour, "An exploratory game-theoretic analysis of biomass electricity generation supply chain," Energy Policy, vol. 37, no. 11, pp. 4514–4522, 2009.

128. M. Mobini, T. Sowlati, and S. Sokhansanj, "Forest biomass supply logistics for a power plant using the discrete-event simulation approach," Applied Energy, vol. 88, no. 4, pp. 1241–1250, 2011.

129. M. Mahnam, M. R. Yadollahpour, V. Famil-Dardashti, and S. R. Hejazi, "Supply chain modeling in uncertain environment with bi-objective approach," Computers and Industrial Engineering, vol. 56, no. 4, pp. 1535–1544, 2009.

130. S. Sokhansanj, A. Turhollow, and E. Wilkerson, "Integrated biomass supply and logistics," Resource, vol. 15, no. 6, pp. 15–18, 2008.

131. M. B. Alam, R. Pulkki, C. Shahi, and T. Upadhyay, "Modeling woody biomass procurement for bioenergy production at the Atikokan Generating Station in Northwestern Ontario, Canada," Energies, vol. 5, no. 12, pp. 5065–5085, 2012.

132. T. P. Upadhyay, C. Shahi, M. Leitch, and R. Pulkki, "Economic feasibility of biomass gasification for power generation in three selected communities of northwestern Ontario, Canada," Energy Policy, vol. 44, pp. 235–244, 2012.

133. J. Nagel, "Determination of an economic energy supply structure based on biomass using a mixed-integer linear optimization model," Ecological Engineering, vol. 16, no. 1, pp. S91–S102, 2000.

134. B. M. Sopha, C. A. Klöckner, and E. G. Hertwich, "Exploring policy options for a transition to sustainable heating system diffusion using an agent-based simulation," Energy Policy, vol. 39, no. 5, pp. 2722–2729, 2011.

135. A. Sultana and A. Kumar, "Optimal siting and size of bioenergy facilities using geographic information system," Applied Energy, vol. 94, pp. 192–201, 2012.

136. H. K. Sjølie, G. S. Latta, D. M. Adams, and B. Solberg, "Impacts of agent information assumptions in forest sector modeling," Journal of Forest Economics, vol. 17, no. 2, pp. 169–184, 2011.

137. M. Aydinel, T. Sowlati, X. Cerda, E. Cope, and M. Gerschman, "Optimization of production allocation and transportation of customer orders for a leading forest products company,"Mathematical and Computer Modelling, vol. 48, no. 7-8, pp. 1158–1169, 2008.

138. H. Gunnarsson, M. Rönnqvist, and J. T. Lundgren, "Supply chain modelling of forest fuel," European Journal of Operational Research, vol. 158, no. 1, pp. 103–123, 2004

139. B. Velazquez-Marti and E. Fernandez-Gonzalez, "Mathematical algorithms to locate factories to transform biomass in bioenergy focused on logistic network construction," Renewable Energy, vol. 35, no. 9, pp. 2136–2142, 2010.

140. H. L. Lam, P. S. Varbanov, and J. J. Klemeš, "Optimisation of regional energy supply chains utilising renewables: p-graph approach," Computers and Chemical Engineering, vol. 34, no. 5, pp. 782–792, 2010.

141. P. C. Jones and J. W. Ohlmann, "Long-range timber supply planning for a vertically integrated paper mill," European Journal of Operational Research, vol. 191, no. 2, pp. 557–570, 2008.

142. D. Carlsson and M. Rönnqvist, "Supply chain management in forestry—case studies at Södra Cell AB,"European Journal of Operational Research, vol. 163, no. 3, pp. 589–616, 2005.

143. S. S. Pitty, W. Li, A. Adhitya, R. Srinivasan, and I. A. Karimi, "Decision support for integrated refinery supply chains. Part 1. Dynamic simulation," Computers and Chemical Engineering, vol. 32, no. 11, pp. 2767–2786, 2008.

144. L. Y. Koo, A. Adhitya, R. Srinivasan, and I. A. Karimi, "Decision support for integrated refinery supply chains. Part 2. Design and operation," Computers and Chemical Engineering, vol. 32, no. 11, pp. 2787–2800, 2008.

145. O. Ahumada and J. R. Villalobos, "Operational model for planning the harvest and distribution of perishable agricultural products," International Journal of Production Economics, vol. 133, no. 2, pp. 677–687, 2011.

146. X. Fang, C. Zhang, D. J. Robb, and J. D. Blackburn, "Decision support for lead time and demand variability reduction," Omega, vol. 41, no. 2, pp. 390–396, 2012.

147. C. A. Garcia, A. Ibeas, J. Herrera, and R. Vilanova, "Inventory control for the supply chain: an adaptive control approach based on the identification of the lead-time," Omega, vol. 40, no. 3, pp. 314–327, 2012.

148. C. Li and S. Liu, "A stochastic network model for ordering analysis in multi-stage supply chain systems," Simulation Modelling Practice and Theory, vol. 22, pp. 92–108, 2012.

149. K. Mitra, R. D. Gudi, S. C. Patwardhan, and G. Sardar, "Towards resilient supply chains: uncertainty analysis using fuzzy mathematical programming," Chemical Engineering Research and Design, vol. 87, no. 7, pp. 967–981, 2009.

150. C. Gomes da Silva, J. Figueira, J. Lisboa, and S. Barman, "An interactive decision support system for an aggregate production planning model based on multiple criteria mixed integer linear programming," Omega, vol. 34, no. 2, pp. 167–177, 2006.

151. P. Luathep, A. Sumalee, W. H. K. Lam, Z.-C. Li, and H. K. Lo, "Global optimization method for mixed transportation network design problem: a mixed-integer linear programming approach," Transportation Research B, vol. 45, no. 5, pp. 808–827, 2011.

152. P. K. Naraharisetti, I. A. Karimi, and R. Srinivasan, "Supply chain redesign through optimal asset management and capital budgeting," Computers and Chemical Engineering, vol. 32, no. 12, pp. 3153–3169, 2008.

153. G. J. Hahn and H. Kuhn, "Designing decision support systems for value-based management: a survey and an architecture," Decision Support Systems, vol. 53, no. 3, pp. 591–598, 2012.

154. A. Sadegheih and P. R. Drake, "System network planning expansion using mathematical programming, genetic algorithms and tabu search," Energy Conversion and Management, vol. 49, no. 6, pp. 1557–1566, 2008.

155. H. J. Ko and G. W. Evans, "A genetic algorithm-based heuristic for the dynamic integrated forward/reverse logistics network for 3PLs," Computers and Operations Research, vol. 34, no. 2, pp. 346–366, 2007.

156. M. Bashiri, H. Badri, and J. Talebi, "A new approach to tactical and strategic planning in production-distribution networks," Applied Mathematical Modelling, vol. 36, no. 4, pp. 1703–1717, 2012.

Global Renewable Energy-Based Electricity Generation and Smart Grid System for Energy Security

M. A. Islam, M. Hasanuzzaman, N. A. Rahim, A. Nahar, and M. Hosenuzzaman

UM Power Energy Dedicated Advanced Centre (UMPEDAC), Level 4, Wisma R&D University of Malaya, Jalan Pantai Baharu, 59990 Kuala Lumpur, Malaysia

ABSTRACT

Energy is an indispensable factor for the economic growth and development of a country. Energy consumption is rapidly increasing worldwide. To fulfill this energy demand, alternative energy sources and efficient utilization are being explored. Various sources of renewable energy and their efficient utilization are comprehensively reviewed and presented in this paper. Also the trend in research and development for the technological advancement of energy utilization and smart grid system for future energy security is presented. Results show that renewable energy resources are becoming more prevalent as more electricity generation becomes necessary and could provide half of the total energy demands by 2050. To satisfy the future energy demand, the smart grid system can be used as an efficient system for energy security. The smart grid also delivers significant environmental benefits by conservation and renewable generation integration.

INTRODUCTION

Economic growth, automation, and modernization mainly depend on the security of energy supply. Global energy demand is rapidly growing, and, presently, the worldwide concern is on how to satisfy the future energy demand. Long-term projections indicate that the energy demand will rapidly increase worldwide. To supply this energy demand, fossil fuels have been used as primary energy sources. Fossil fuels emit greenhouse gases that highly affect the environment and the future generation [1–6]. The emissions largely depend on the emission factor of primary energy sources (i.e., input fuel of the plant). Among all energy sources, the emission factor of fossil fuels (i.e., coal, natural gas, and oil) is very high, as shown in Table 1. Fossil fuels are widely used as the main fuel in power generation. In Malaysia, fossil fuels (i.e., natural gas [53.3%] and coal [26.3%]) serve as major power generation sources, as shown in Figure 1. Large-scale use of fossil fuels, however, greatly affects the environment. Based on the global CO_2 distribution in 2013, the emission breakdown is as follows: coal (43%), oil (33%), gas (18%), cement (5.3%), and gas flare (0.6%) [7].

Table 1: Emission factors of fossil fuels for electricity generation [84]

Fuel	Emission factor (kg/kWh)		
	CO_2	SO_2	NO_x
Coal	1.1800	0.019	0.0052
Petroleum	0.8500	0.0164	0.0025
Gas	0.5300	0.0005	0.0009

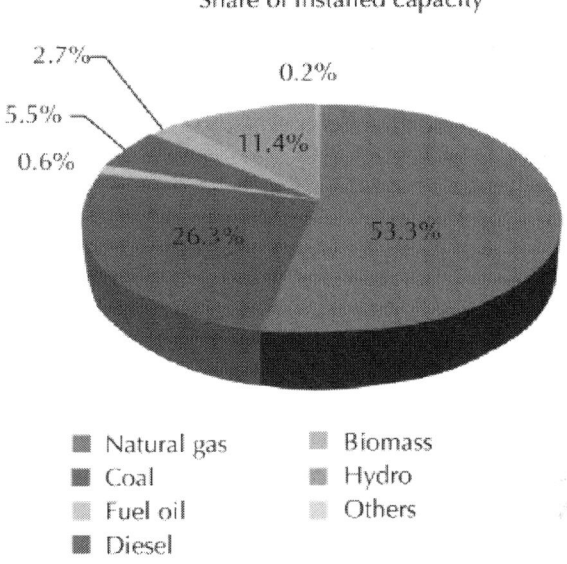

2.7%   0.2%

5.5%

0.6%   11.4%

26.3%   53.3%

■ Natural gas   ■ Biomass
■ Coal   ■ Hydro
■ Fuel oil   ■ Others
■ Diesel

Figure 1: Share of installed capacity as of December 31, 2012, in Malaysia [88].

Meanwhile, renewable energy sources (solar, wind, hydro, geothermal, biomass, etc.) are emission-free energy sources in the world. Renewable energy technologies are an ideal solution because they can contribute significantly to worldwide power production with less emission of greenhouse gases [8–11]. The "sustainable future" scenario of the International Energy Agency (IEA) shows 57% of world electricity being provided by renewable energy sources by 2050 [12]. Long-term forecast and planning is required to achieve this ultimate target [9]. Renewable energy-based power generation and supply to the national grid for a specific zone are necessary. The conventional grid aggregates the multiple networks, and the regulation system consists of various levels of communication and coordination, in which most of the systems are manually controlled [13]. A smart grid is a new concept that leads to the modernization of the transmission and distribution grid. The smart grid system is the digital upgrade of transmission and new markets for the alternative energy generation of renewable energy sources. Presently, smart grid is an often-cited term in the energy generation and distribution industry [14].

Smart grid connected with distributed power generation is a new platform that significantly generates reliable security of supply (SOS) and quality of electric energy. This concept is practical and reliable as numerous types of energy sources become available, such as solar, wind, biomass, and hydropower. Renewable and nonconventional energy sources are allowed to integrate with the distributed power generation link that has a smart grid. This study therefore highlights the role of renewable energy sources in generating electricity and the integration with the smart grid system for energy security.

WORLD ENERGY CONSUMPTION SCENARIO

Global population growth and improvement of living standards cause high energy demand [15]. This global energy demand is increasing faster than the population growth rate [16]. Approximately 80% of the total primary energy is being supplied by fossil fuels [17]. World energy consumption projection from 2002 to 2030 shows the increase of energy demand by almost 60% (1.6% per year). The energy demand will be approximately 16.5 billion tons of oil equivalents (toe) by 2030 compared with 10.3 billion in 2002 [18, 19]. Tables 2 and 3present the projection of the world primary energy demand from 2002 to 2030

Table 2: World primary energy demand (Mtoe) [18]

Energy sources	2002	2010	2020	2030	2002–2030
Coal	2389	2763	3193	3601	1.50%
Oil	3676	4308	5074	5766	1.60%
Gas	2190	2703	3451	4130	2.30%
Nuclear	692	778	776	764	0.40%
Hydro	224	276	321	365	1.80%
Biomass and waste	1119	1264	1428	1605	1.30%
Others renewable	55	101	162	256	5.70%
	10345	12193	14405	16487	1.70%

Table 3: World primary energy demand (%) from 2002 to 2030 [18]

Energy sources	2002	2010	**2020**	**2030**
Coal	23%	23%	22%	22%
Oil	36%	35%	35%	35%
Gas	21%	22%	24%	25%
Nuclear	7%	6%	5%	5%
Hydro	2%	2%	2%	2%
Biomass and waste	11%	10%	10%	10%
Others renewable	0.53%	0.83%	1.12%	1.55%

The global primary energy consumption projection shows that fossil fuels will solely contribute the largest amount of energy, not taking into consideration if its share will slightly decrease from 36% in 2002 to 35% in 2030 [20]. The demand for gas in the power sector will tremendously increase to a maximum of 60% by 2030, and the market share will increase up to 47% by 2030 compared with 36% in 2002. In most parts of the world, natural gas is expected to remain as a main competitive fuel in the new power station because of its high efficiency [21–23]. The share of coal in fulfilling the total primary energy demand will decrease from 30% in 2012 to 27% by 2035 [24]. The nuclear energy will decrease to 5% by 2030 from 7% in 2002. The intensities in different regions will continue to vary because of levels of variation in economic growth, energy use of different users, energy prices, geography, economic arrangement, culture, lifestyle, and climate [19, 21–23].

SOURCES OF RENEWABLE ENERGY FOR A GREEN AND CLEAN WORLD

Adequate energy sources and supply are essential for the economic and social growth of any nation. The development of the energy supply system is the main objective of the United Nations Millennium Declaration. In developed countries, innovative technologies and modern concepts have to be upgraded to enhance energy efficiency, which is considered as the performance indicator of the millennium declaration. From an international perspective, renewable energies provide several benefits to the conventional energy system. Renewable energy reduces CO_2 emissions, which is the main aim of climate protection. Renewable energy reduces dependency on fossil fuels and energy import from other countries and improves economic growth [18, 25–31]. Presently, enormous challenges are being encountered regarding the necessities of life. Moreover, service requirements continuously increase with the increase of power consumption. According to IEA [32] in Paris, by 2030, approximately 60% of energy utilization will increase compared with the utilization level in 2001. Fulfilling the energy demand by using fossil fuels as primary resource would be more difficult. Power generation from fossil fuels can potentially harm the environment and can cause global warming. Thus, the next-generation energy system must be sustainable and carbon-free. Policy makers should promote renewable energy (i.e., solar, wind, biomass, hydropower, and geothermal) as primary sources of energy. The potentiality of the current renewable energy technology is extensive and positive. Review of the literature shows that half of the total energy demands could be satisfied by renewable sources by 2050 [25, 33, 34]. The share of renewable energy sources is predicted to contribute approximately 30% to 80% in 2100 [35]. The total energy share of various fields of energy sources and particularly the contribution of renewable energy (16.70%) sources of the total world energy consumption is shown in Figure 2. The annual average growth rate of the capacity of the world renewable energy between 2006 and 2011 is shown in Figure 3. Table 4 shows the summary of world renewable energy use by type and scenario. The renewable energy-based electricity generation is increasing; that is, the share of total electricity generation was 20% in 2010 and will be 31% by 2035. Figure 4 shows that the rate of renewable energy-based generation in Organization for Economic Cooperation and Development (OECD) is higher (approximately 18% in 2010 and 33% by 2035) than that in other regions [36].

Table 4: World renewable energy use by type and scenario [85]

	New policies			Current policies		450 scenario	
	2011	2020	2035	2020	2035	2020	2035
Electricity generation (TWh)	4 482	7 196	11 612	6 844	10 022	7 528	15 483
Bioenergy	424	762	1 477	734	1 250	797	2 056
Hydro	3 490	4 555	5 827	4 412	5 478	4 667	6 394
Wind	434	1 326	2 774	1 195	2 251	1 441	4 337
Geothermal	69	128	299	114	217	142	436
Solar PV	61	379	951	352	680	422	1 389
Concentrating solar power	2	43	245	35	122	56 806	56 806
Marine	1	3	39	3	24	3	64
Share of total generation	20%	26%	31%	24%	25%	28%	48%
Heat demand* (Mtoe)	434	438	602	432	551	466	704
Industry	209	253	316	255	308	248	328
Buildings* and agriculture	135	184	286	177	243	198	376
Share of total demand	8%	10%	12%	9%	11%	10%	16%
Biofuels (mboe/d)**	1.3	2.1	4.1	1.9	3.3	2.6	7.7
Road transport	1.3	2.1	4.1	1.9	3.2	2.6	6.8
Aviation***	—	—	0.1	—	0.1	—	0.9
Share of total transport	2%	4%	6%	3%	4%	5%	15%
Traditional biomass (Mtoe)	744	730	680	732	689	718	647
Share of total bioenergy	57%	49%	37%	50%	40%	47%	29%
Share of total renewable energy demand	43%	33%	22%	34%	25%	32%	17%

*Excluding traditional biomass. **Expressed in energy-equivalent volumes of gasoline and diesel.

***International bunkers. Note: Mtoe: million tonnes of oil equivalent; TPED: total primary energy demand; TWh: terawatt-hour; mboe/d: million barrels of oil equivalent per day.

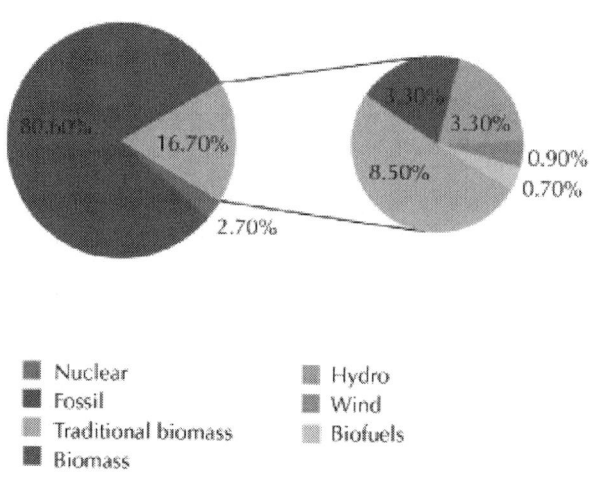

- Nuclear
- Fossil
- Traditional biomass
- Biomass
- Hydro
- Wind
- Biofuels

Figure 2: Global energy consumption and share of renewable energy, 2010 [89].

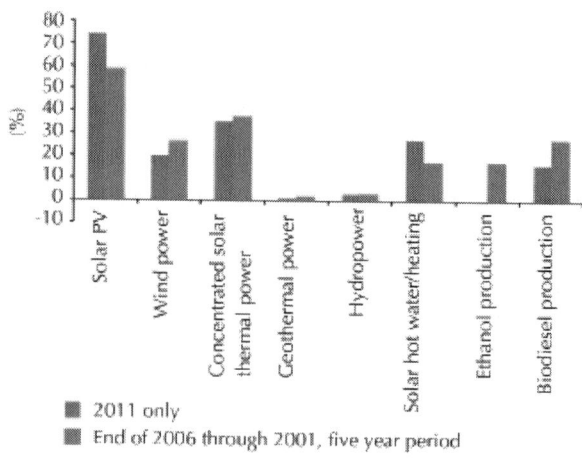

Figure 3: Renewable energy capacity growth rate, 2006–2011 [89].

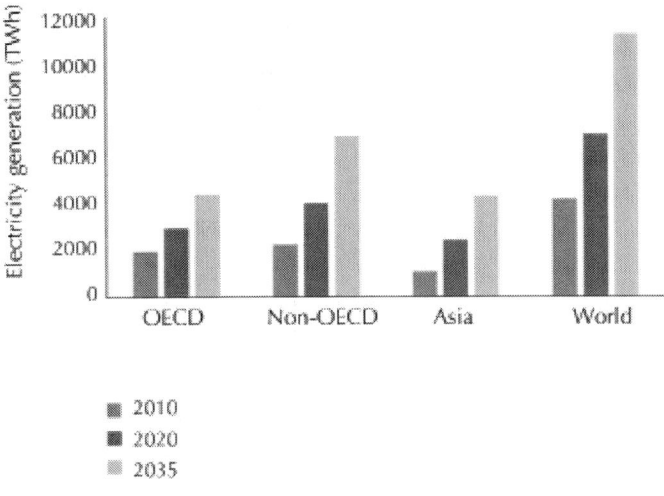

Figure 4: Renewable-based electricity generation by region [36].

The renewable energy-based electricity generation has been increasing where the share of total electricity generation was 20% in 2010 and will be 31% by 2035. Figure 4 shows that the rate of renewable energy-based generation in OECD is higher (about 18% in 2010 and 33% by 2035) than other regions [36].

SMART GRID AND ENERGY SECURITY

To improve the sturdiness and reliability of the grid, the security systems should be improved in both the physical and cyber perspectives. Ultimately, this action will reduce the probability and consequences of man-made occurrences. Energy security is a concept that ensures the reliability of energy sources, maintains a sufficient energy supply at an affordable price, and prevents the harmful effects to the environment. Energy security is a multidimensional issue that addresses risk management, diversity of energy, and decision making for implementing the policy [37]. The integration of renewable energy using smart grid technologies can improve energy security and safety of the electric system.

Smart Grid

A smart grid is the solution to the modernization of the electrical energy system and infrastructure to present a more intelligent and reliable electricity grid. Smart grids provide many benefits over conventional grid. Smart grids improve both the physical and economic operations of the grid system, increasing reliability and sustainability [38]. A more conceptual definition of smart grid is presented by Rahman [39]. According to the modern technology-based grid initiative of the United States Department of Energy, an intelligent self-response is based on the demand or a smart grid integrating and combining with advanced sensing, monitoring technologies control methods, and two-way communications into the current electricity grid. Figure 5 shows the block diagram of the smart grid concept. Energy security and optimization of the demand can be minimized by implementing the smart grid system.

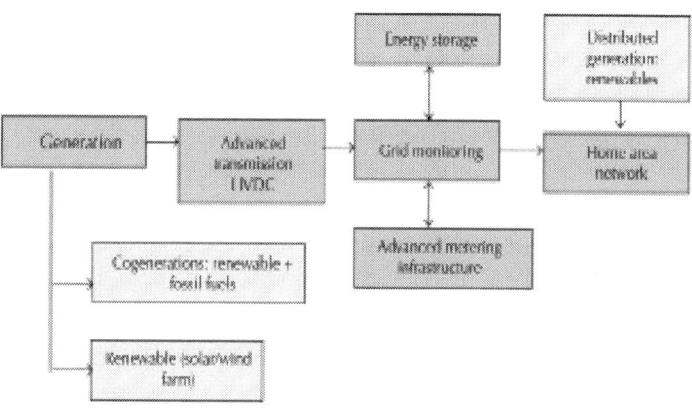

Figure 5: Block diagram of the smart grid concept [86].

The smart grid system is designed to handle uncertain incidents. The three security objectives of smart grid are to ensure (1) the availability of the power supply based on customer requirement, (2) the two-way communication system, and (3) the data security of the customer [40]. The smart grid mainly aims to enhance overall management, which refers to obtaining better control of the transmission system that will improve system reliability. This technique has numerous advantages with regard to frugality, despite the low energy efficiencies (system losses occurring along the distribution line). Smart grid technologies are capable of supporting the system operator in controlling and managing the energy streams on the grid with more accuracy by applying the flexible AC transmission systems. First, using a modern sensor that is called a phasor measurement unit that determines the real-time response of service providers, the efficiency of the total electricity system is improved [41]. Second, the automation of the smart grid will be more self-responsive, and better control of the substation on the distributed network is ensured. The distribution system automation of the smart grid allows utility firms to upsurge the strong communication of the distribution network and prevents the interruption of supply to the end user at unanticipated incidents such as an environmental hazard that destroys power poles or causes infrastructure damage to the substation. The end-user load is also controlled by implementing the distribution channel automation. The integration of modern communication technology with various grid segments provides better and reliable service to the end user, which is the basic role of the smart grid. Figure 6 shows the comparison between conventional and smart grids.

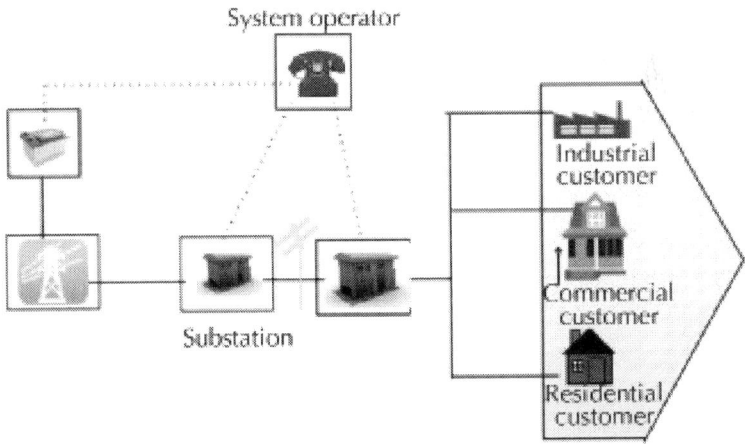

Figure 6: The comparison between conventional grid and smart grid [90].

Figure 7 shows the smart grid domains through secure communication and electrical flows. This control system logically increases the confidential issues of individual end-user level information [42]. The peak load time of end-user appliances is automatically recognized by a smart metering system, without employing a service person to collect the data from the electric meter.

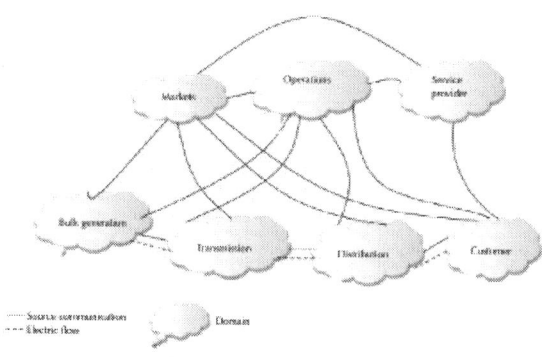

Figure 7: NIST smart grid domains through secure communication flows and electrical flows [91].

Smart Grid Technology and Applications

Modern engineering tools and techniques are required to develop the smart grid. The integration of information technology, strong monitoring system, and practical strategic plan is necessary to completely understand the smart grid application. The demand for electricity being satisfied by the centralized and distributed generation (DG) system through the use of a smart grid technology is a very modern and reliable concept. The system operation and control system of the smart grid are monitored by modern information and communication technologies that enable the operator to practice control over the demand and efficiently provide reliable and high-quality service. The smart grid provides the most effective electrical distribution network through the two-way communication system based on the responses of the customer. Power industries worldwide are unpredictably facing huge challenges. Existing grids are also challenged to perform safely and provide reliable supply. In addition, social and political gain is important and depends on the electricity generation and utilization and its environmental impacts [43]. Developing countries are formulating their policy based on the requirement of an enhanced smart grid. A huge amount of federal funding has been allocated to promote and assist the smart grid policy in different states. Better management for the smart grid in the electricity industry is time-consuming. The smart grid is integrated into the infrastructure that supplies electricity, which is coupled with modern telecommunication, IT, and sensing technology. The great potentiality of the smart grid is defined by its capability to process and to analyze a huge amount of data and to implement critical demand management. The smart grid provides flexible opportunity to the system operator and the end users with its use of artificial intelligence and integration with the computer system. The application of the smart grid in developed countries is happening much quicker than its display of numerous benefits [44]. The end user is the dynamic player in the electricity industry. Benefits and savings can be attained by optimizing peaks in demand and by increasing the energy performance. The successful application of smart grid is the key factor in reaching its ultimate aims to reduce greenhouse gas discharges and to utilize energy efficiently [45]. Table 5 shows the smart grid technologies, applications, and purposes. Demand management would be quicker if the network would connect with the innovation, new energy products, and services. As a result, the information channel, statistics storing and management, and rules of governing access by various customers are developed. The eight priority areas in building a smart grid (as identified by the National Institute of Standards and Technology (NIST)) are as follows [46]: (a) demand response and consumer energy efficiency, (b) wide-area situational awareness, (c) energy storage, (d) electric transportation, (e) advanced metering infrastructure, (f) distribution grid management, (g) cyber security, and (h) network communications.

Table 5: Smart grid technologies, applications, and purposes [86]

Application	Technology	Purpose
Distributed automation	Alternate energy Smart sensing Advanced smart metering	Reduces system losses
Data analysis	Information technology	Collect and analyze from the grid
Demand response	Smart appliances	To achieve lower electricity rates
Carbon management	Alternate energy Smart sensing Advanced smart metering	Reduce carbon footprint
Home energy management	Smart sensing Advanced smart metering Smart appliances	Track and optimize energy use

Reliability of the Smart Grid

Reliability issues in modern power grids are becoming more challenging. The challenges include aggravated grid congestion, larger transfers over longer distances, increasing volatility, and reduced reliability margins [47]. The integrated network of islanding connected with DG could increase reliability and enhance the good quality service of the local electricity supply. The electrical charge based on real-time sensing is practical for the consumer; this charge is possible by implementing advance metering [48]. The smart grid possesses particular characteristics or delivers the following: it self-heals from power disturbance events, permits end-user participation in demand management, robustly manages physical damage and cyber-attacks, provides the power quality required in the 21st century, is flexible in accepting all generation and storage options, permits the introduction of new products, services, and markets, and develops the operating efficiency and optimization of assets [41]. Smart meters now account for 85% of all such devices in Italy and 25% in France. Many governments have aimed at nationwide placements of smart grids by 2020 [49]. Developed countries also envision the amplification and application of energy or climate protection policies and solve problems relevant to them [50]. Each country has its specific viewpoint in gaining the benefits of the market segment from smart grids.

Energy Security

The focus on energy security and reliability is based on the concept that a continuous supply of energy is critical for an effective economy of any country. The meaning of security of energy or SOS varies among different people at different places [51]. The SOS of energy is mainly associated with the security of access to oil or gas supplies and is connected with the future scenario of maintaining the fossil fuel reserve of a particular nation. Every day the definition and concept of energy security change and are modified by the changes in technology. Presently, energy security is defined by four main and basic elements [52–55]. Security of energy involves the readiness or real existence of fossil fuels, gap, and discrepancy between energy consumption and generation, cost involved with the SOS, and environmental sustainability (e.g., related to the obtainability of solar, wind, and bioenergy). The Asia Pacific Energy Research Center [54] has identified and classified the elements that are related to SOS as follows:(i)accessibility or geopolitical elements;(ii)acceptability or environmental and societal elements;(iii) availability or elements relating to geological existence;(iv)affordability or economical elements.

Indicators of Energy Security

Energy security is an important factor for economic development and consists of several relevant factors that are aggregated using variables. Energy security indicators explain the contribution of SOS. Most of the indicators are subjected to a particular context. An individual indicator has particular significance. Thus, understanding the suitability and application of dissimilar SOS indicators is important. SOS indicators are used in analyzing the scenario of energy SOS under different perspectives [56].

Energy Resources and Import Dependency

Energy reserve, where energy resources are considered as direct indicators, is vital for SOS [56]. Reduction and adjustment of energy demand can contribute to energy security. Generally, residential and business areas with tolerable demands are easier to supply and are less exposed to energy price increases. Thus, the tolerable energy demand of any place can cause few opportunities for energy to be imported from other countries. The demand for energy involves the following: energy demand per home or unit of economic activity, energy costs as a proportion of total expenditure that indicates the severity of exposure to price increases, and capacity for demand-side response [57]. The existence of hydrocarbon resources is very uncertain and unpredictable, and modern technology is necessary to explore these resources. The United States Geological Survey [58] is one of the famous organizations that provide the most authentic information relevant to geographical resource estimation. The data provided by this organization are considered as the most recognized, reliable, and autonomous [59]. The most frequently used SOS indicator is import dependence. This indicator refers to the importation of oil as being dependent on other regions and is often relative to oil consumption [60]. This indicator is commonly used in measuring energy security; that is, high import dependence means low energy security [56].

Indices of Energy Diversity

Diversity of energy lessens the probability of disruption of energy supply. However, how an energy system may be diversified and how it can be measured remain unclear. Diversity index is a combination of several concepts, such as energy resources used by an economic sector or country, countries and/or companies supplying those resources, and technologies and infrastructure used to convert, transport, and deliver energy to consumers [57]. Diversity of energy sources in various geographical locations is considered as hindrance to supply risks [61, 62]. A statistical quantity of the multiplicity of energy sources is an indicator of SOS. The indicator of diversity consists of three basic factors [63]: variety (number of sets), balance (range across types), and disparity (individuality that makes groups different from one another). However, no index can measure appropriate diversity. Diversity index means measuring the multiplicity or diversity in energy supply, not taking into consideration if the classification is still influenced by the individual option. This formal indicator is not considered as the thread of trouble created by various fuels [56].

Political Stability

The political situation of the country that supplies energy is an important factor in the security of energy supply. Energy supply actually depends on the negotiation between the governing body of the country and other parties. The political risk indicators are mentioned by the World Bank, such as political stability, absence of violence, and regulatory quality [18, 53, 56, 64]. The political indicator is a geopolitical market concentration risk. Political factors associated with the exporting countries are measured by the political indicator. Low value of geopolitical market concentration indicates high political risk [65].

Price of Energy

The demand and supply balancing mechanism is a common function and is influenced by market prices. Prices are affected by the supply in relation to demand; thus they are recognized as a measure of economic impacts. The oil price is an important indicator of SOS because it is affected by many factors (i.e., speculation, strategic communication, and short-term shortages) in the market [56]. The lack of supply affects the market price of energy, and the responses of customers contribute to the market equilibrium condition. High price of energy leads consumers to reduce their consumption and to search for alternative sources of energy. The price of energy also involves energy security concerns [66].

Market Liquidity

The ratio of the world oil export to the net oil import of the country is called market liquidity. Market liquidity determines the oil availability in the world market against the portion of oil demand in the domestic market that cannot be supplied by the local oil production. Thus, high market liquidity expresses the necessity for extra oil supply from another country [65]. A close relation exists between market liquidity and SOS in balancing the fluctuation of demand and supply of fuels in the market. IEA stated that [62] market liquidity is the exponential function of the ratio of the consumption of a country over the total fuel offered in the market [56, 67].

SMART GRID AND INTEGRATION OF RENEWABLE ENERGY SOURCES

Given the rising energy prices and the greenhouse effect, renewable resources are more environmentally convenient and more efficient. Solar technology is the most ideal solution to energy demand management and prevention of greenhouse gas emission and is a milestone to the generation of green and clean energy. The most remarkable technologies involve generation techniques that use wind turbines, solar energy, hydropower, and biomass [68–70]. Figures 8, 9, 10, and 11 illustrate the renewable energy-based integration to the grid.

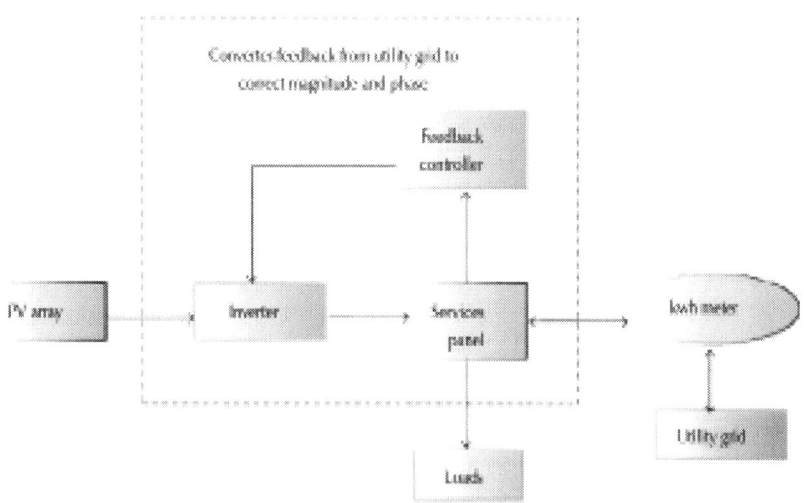

Figure 8: PV-based solar energy integration [92].

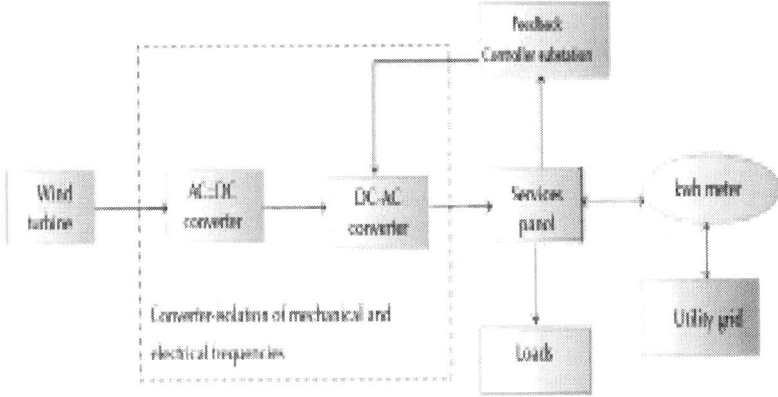

Figure 9: Wind energy integration [92].

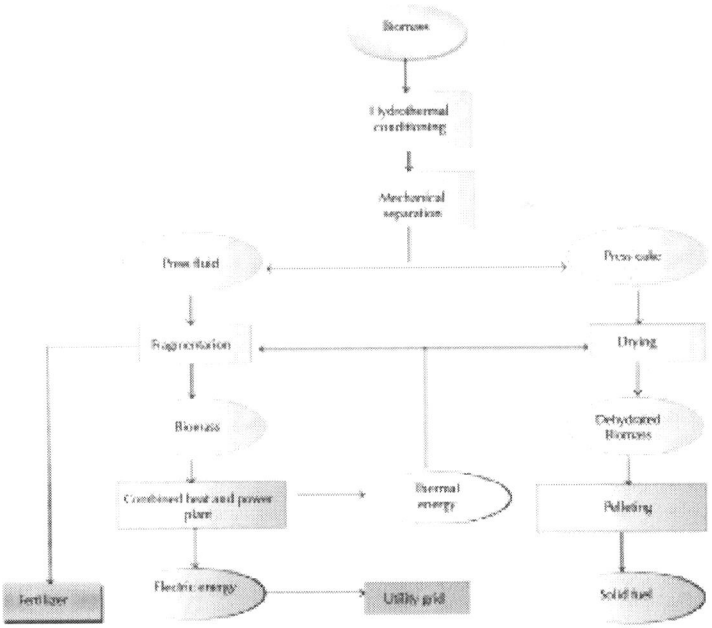

Figure 10: Biomass-based energy [93].

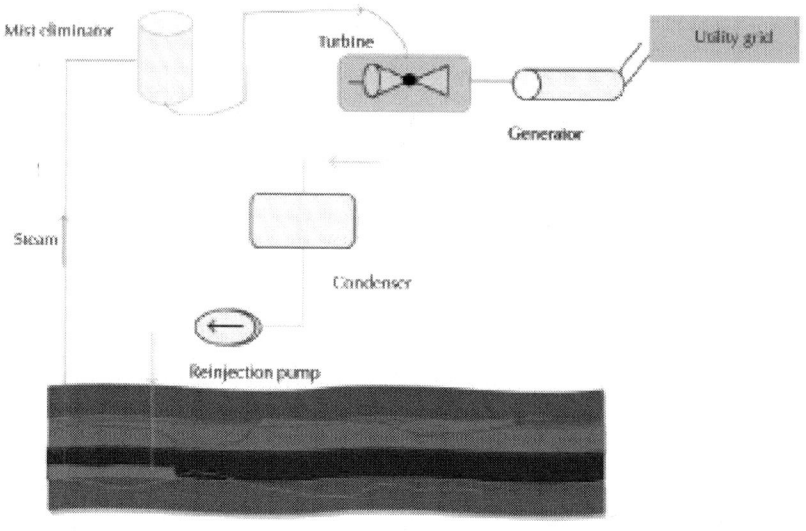

Figure 11: Geothermal-based power generation [94].

The extended application of these types of innovation and technologies depends on three crucial factors that are related to the future energy supply and that introduce the multiapplication of grids: DG, distributed energy storage (DES), and demand-side load management (DSLM). In DG, various energy sources are connected to the power grid. These types of sources range from high- to local-level generators such as combined or hybrid power plants [68, 71].

Local power generation using renewable energy is vital in the implementation of the smart grid [72]. The handling of renewable energy sources requires refined arrangement and operation planning based on the concept of technologies [73]. The large-scale distributed renewable generation system requires a more flexible, reliable, and smarter grid [74]. Energy storage is also the fundamental concept of the smart grid. This system provides support to the development of sustainable energy [75]. Figure 12 shows the overall renewable energy integration and storage system. Given the growing participation of renewable energy to the electricity supply chain, the necessity of energy storage will ultimately increase [68]. The application of a large amount of electricity power storage is challenging and can cause system losses. The application of DES is a possible solution because it slightly reduces the disadvantages of the energy backup requirements of the smart grid. The other benefit of the DES application in a smart grid is the enhancement of DSLM by a small-scale backup policy [76] and the improvement of the generation performance by supporting the peak demand [77, 78]. The local backup of energy is used for the efficient management of addressing energy demand during the peak period. The management of peak demand by efficient and bright management techniques will make the grid reliable and smarter and perform better [75, 76]. Using a hybrid system of power generation will result in a more balanced and controlled management of the grid. As the power generation becomes close to the end user, the hybrid power generation technology provides the service; thus no transmission losses exist [79]. Hybrid generation connected with the smart grid approach has numerous advantages over the conventional system. First, the cost of transmission and distribution is low. Approximately 30% of the cost is involved with electric transmission to other places in a conventional system. The domestic supply line does not have high capital for the initial setup and has less energy losses from long-distance transmission lines. Local connection lines do not have high capital costs and energy losses caused by long-distance distribution lines; energy consumption also decreases because of new innovative devices [80]. Second, DG allows the integration of local renewable energy sources to power plants and is helpful in reducing greenhouse gas emission [81]. The development of hybrid power generation connected with the smart grid reduces emission and satisfies the level of electricity supply against end-user demand [82, 83]. Table 6 shows the potential reductions in electricity and CO_2 emissions in 2030 attributable to smart grid technologies.

Table 6: Potential reductions in electricity and CO_2 emissions in 2030 attributable to smart grid technologies [87]

Mechanism	Reductions in electricity sector energy and CO_2 emissions*	
	Direct (%)	Indirect (%)
User information and feedback systems	3	—
Categorization of residential and small/medium commercial buildings	3	—
Energy efficiency measurement and verification programs	1	0.5
Shifting load	<0.1	—
Support electric vehicle and plug-in hybrid electric vehicle	3	—
Advanced voltage control	3	—
Support penetration of renewable energy generation (25% renewable portfolio standard)	<0.1	5
Total reduction	12	6

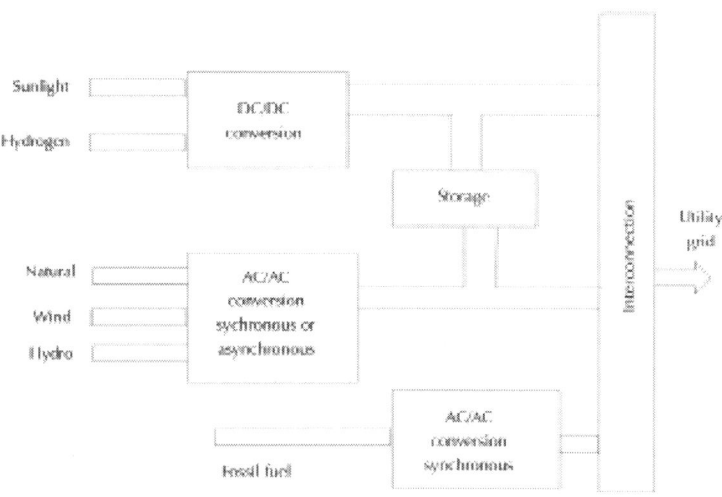

Figure 12: Integration of several sources of energy into the grid.

CONCLUSIONS

Ensuring a reliable, efficient, and affordable energy is a great challenge. Generating electricity from renewable energy sources can provide direct and indirect economic benefits in excess of costs as well as environmental benefits through the reduction of CO_2 emission. Policy makers should promote renewable resources (i.e., solar, wind, biomass, hydropower, and geothermal) for sustainable and carbon-free energy. It is predicted that about 57% of total energy demand could be generated from renewable sources by 2050. The renewable energy source power generation integrated into the smart grid system can be one of the best options for future energy security. The smart grid system addresses the degradation of energy source and modern information technology for communication and improves the efficiency of power distribution. A smart grid can transform the 20th century power grid as a more intelligent, flexible, reliable, self-balancing, and interactive network that enables economic growth, environmental oversight, operational efficiency, energy security, and increased consumer control. Moreover, the smart grid would create new markets as private industries develop energy-efficient and intelligent appliances, new communication capabilities, and smart meters. Smart grid can replace traditional forms of energy with renewable sources of generation. Renewable energy is always required by environmentalists in the hopes of developing a cleaner and more efficient power generation. A smart grid is environmentally beneficial because it utilizes the distribution of renewable sources. Smart grid offers a genuine path toward significant environmental improvement.

ACKNOWLEDGMENTS

The authors would like to acknowledge the financial support from the High Impact Research Grant (HIRG) scheme (Project no. UM.C/HIR/MOHE/H-16001-00-D000032, Campus Network Smart-Grid System for Energy Security) to carry out this research.

REFERENCES

1. M. Hasanuzzaman, N. A. Rahim, M. Hosenuzzaman, R. Saidur, I. M. Mahbubul, and M. M. Rashid, "Energy savings in the combustion based process heating in industrial sector," Renewable & Sustainable Energy Reviews, vol. 16, no. 7, pp. 4527–4536, 2012. · ·

2. M. Hasanuzzaman, N. A. Rahim, R. Saidur, and S. N. Kazi, "Energy savings and emissions reductions for rewinding and replacement of industrial motor," Energy, vol. 36, no. 1, pp. 233–240, 2011.

3. J. Zhang, M. Fu, Y. Geng, and J. Tao, "Energy saving and emission reduction: a project of coal-resource integration in Shanxi Province, China," Energy Policy, vol. 39, no. 6, pp. 3029–3032, 2011.

4. J. M. Joelsson and L. Gustavsson, "Reduction of CO_2 emission and oil dependency with biomass-based polygeneration," Biomass and Bioenergy, vol. 34, no. 7, pp. 967–984, 2010

5. Z. M. Chen and G. Q. Chen, "An overview of energy consumption of the globalized world economy,"Energy Policy, vol. 39, no. 10, pp. 5920–5928, 2011.

6. M. A. Khan, M. Z. Khan, K. Zaman, and L. Naz, "Global estimates of energy consumption and greenhouse gas emissions," Renewable and Sustainable Energy Reviews, vol. 29, pp. 336–344, 2014.

7. GCE, "Global Carbon Emissions—CO2 Now—Current CO2," http://co2now.org/, 2013.

8. P. E. Ugwuoke, U. C. Agwunobi, and A. O. Aliyu, "Renewable energy as a climate change mitigation strategy in Nigeria," International Journal of Environmental Sciences, vol. 3, no. 1, pp. 11–19, 2012.

9. D. N. Nkwetta, M. Smyth, A. Zacharopoulos, and T. Hyde, "Optical evaluation and analysis of an internal low-concentrated evacuated tube heat pipe solar collector for powering solar air-conditioning systems,"Renewable Energy, vol. 39, no. 1, pp. 65–70, 2012. · ·

10. DTI (Department of Trade and Industry), Energy White Paper: Our Energy Future-Creating a Low Carbon Economy, The Stationary Office, London, UK, 2003.

11. F. Ahmed, A. Q. Al Amin, M. Hasanuzzaman, and R. Saidur, "Alternative energy resources in Bangladesh and future prospect," Renewable and Sustainable Energy Reviews, vol. 25, pp. 698–707, 2013. · ·

12. IEA, Renewable energy outlook/Global Energy Trends, 2012.

13. T. Vijayapriya and D. P. Kothari, "Smart grid: an overview," Smart Grid and Renewable Energy, vol. 2, pp. 305–311, 2011.

14. M. Wissner, "The Smart Grid—a saucerful of secrets?" Applied Energy, vol. 88, no. 7, pp. 2509–2518, 2011. · ·

15. D. Markovic, B. Ceperkovic, and A. Vlajcic, The White Book of the Electric Power Industry of Serbia, PE Electric Power Industry of Serbia, Belgrade, Serbia, 2011.

16. M. H. Hasan, T. M. I. Mahlia, and H. Nur, "A review on energy scenario and sustainable energy in Indonesia," Renewable and Sustainable Energy Reviews, vol. 16, no. 4, pp. 2316–2328, 2012.

17. A. Ummadisingu and M. S. Soni, "Concentrating solar power: technology, potential and policy in India,"Renewable and Sustainable Energy Reviews, vol. 15, no. 9, pp. 5169–5175, 2011

18. International Energy Agency (IEA), World Energy Outlook 2004, OECD/IEA, Paris, France, 2004.

19. K. Bilen, O. Ozyurt, K. Bakirci et al., "Energy production, consumption, and environmental pollution for sustainable development: a case study in Turkey," Renewable and Sustainable Energy Reviews, vol. 12, no. 6, pp. 1529–1561, 2008.

20. ExxonMobil, "2012 The Outlook for Energy: A View to 2040," 2012.

21. BP, "British Petroleum (BP)," BP, London, UK, 2004.

22. Goldemberg, World energy assessment overview, 2004.

23. IEA, "International Energy Agency (IEA)," OECD/IEA, Paris, France, 2000.

24. BP, "BP Energy Outlook 2035," 2014.

25. E. Martinot, A. Chaurey, D. Lew, J. R. Moreira, and N. Wamukonya, "Renewable energy markets in developing countries," Annual Review of Energy and the Environment, vol. 27, pp. 309–348, 2002.

26. International Energy Agency (IEA), World Energy Investment Outlook: 2003 Insights, OECD/IEA, Paris, France, 2003.

27. L. B. Becker, Renewable Energy Policies and Barriers, vol. 5 of Encyclopedia of Energy, Academic/Elsevier, New York, NY, USA, 2004.

28. O. Afshar, R. Saidur, M. Hasanuzzaman, and M. Jameel, "A review of thermodynamics and heat transfer in solar refrigeration system," Renewable and Sustainable Energy Reviews, vol. 16, no. 8, pp. 5639–5648, 2012

29. R. Saidur, M. Rezaei, W. K. Muzammil, M. H. Hassan, S. Paria, and M. Hasanuzzaman, "Technologies to recover exhaust heat from internal combustion engines," Renewable and Sustainable Energy Reviews, vol. 16, no. 8, pp. 5649–5659, 2012

30. M. Hasanuzzaman, R. Saidur, and N. A. Rahim, "Analysis of energy and exergy of an annealing furnace," Applied Mechanics and Materials, vol. 110–116, pp. 2156–2162, 2012

31. M. Hasanuzzaman, R. Saidur, and H. H. Masjuki, "Effects of different variables on moisture transfer of household refrigerator-freezer," Energy Education Science and Technology. Part A. Energy Science and Research, vol. 27, no. 2, pp. 401–418, 2011

32. IEA, "World energy outlook," Medium-Term Oil and Gas Market Reports, 2009.

33. Becker, Renewable Energy Policies and Barriers, vol. 5 of Encyclopedia of Energy, Academic/Elsevier, New York, NY, USA, 2004.

34. Energy Information Administration (EIA), International Energy Outlook, 2006.

35. N. L. Panwar, S. C. Kaushik, and S. Kothari, "Role of renewable energy sources in environmental protection: a review," Renewable and Sustainable Energy Reviews, vol. 15, no. 3, pp. 1513–1524, 2011

36. REO, "Renewable energy outlook," World Energy Outlook 2012, 2012,http://www.worldenergyoutlook.org/media/weowebsite/2012/WEO2012_Renewables.pdf.

37. M. C. Chuang and H. W. Ma, "An assessment of Taiwan›s energy policy using multi-dimensional energy security indicators," Renewable and Sustainable Energy Reviews, vol. 17, pp. 301–311, 2013

38. G. P. J. Verbong, S. Beemsterboer, and F. Sengers, "Smart grids or smart users? Involving users in developing a low carbon electricity economy," Energy Policy, vol. 52, pp. 117–125, 2013. · ·

39. S. Rahman, Smart Grid Opportunies and Challenges, IEE Power Energy Socity, 2012.

40. F. Aloula, A. R. Al-Ali, R. Al-Dalky, M. Al-Mardinia, and E. H. Wassim, "Smart grid security: threats, vulnerabilities and solutions," International Journal of Smart Grid and Clean Energy, vol. 1, no. 1, pp. 1–6, 2012.

41. S. Blumsack and A. Fernandez, "Ready or not, here comes the smart grid!," Energy, vol. 37, no. 1, pp. 61–68, 2012

42. P. McDaniel and S. McLaughlin, "Security and privacy challenges in the smart grid," IEEE Security and Privacy, vol. 7, no. 3, pp. 75–77, 2009

43. M. G. Morgan, J. Apt, L. B. Lave, J. Bergerson, and S. Blumsack, "The U.S. electric power sector and climate change mitigation for the pew center on global climate change," Research Showcase @ CMU, pp. 1–84, 2005.

44. I. J. Pérez-Arriaga, "Regulatory instruments for deployment of clean energy technologies," Loyola de Palacio Programmeon Energy Policy, EUIRSCAS, 2010.

45. A. A. El-Sebaii and H. Al-Snani, "Effect of selective coating on thermal performance of flat plate solar air heaters," Energy, vol. 35, no. 4, pp. 1820–1828, 2010. · ·

46. D. H. Mohsenian-Rad, Communications and Control in Smart Grid, 2012.

47. K. Moslehi and R. Kumar, "Smart grid—a reliability perspective," in Proceedings of the Innovative Smart Grid Technologies Conference (ISGT ‹10), IEEE, Washington, DC, USA, January 2010. · ·

48. J. Kiviluoma and P. Meibom, "Methodology for modelling plug-in electric vehicles in the power system and cost estimates for a system with either smart or dumb electric vehicles," Energy, vol. 36, no. 3, pp. 1758–1767, 2011. · ·

49. A. Faruqui, D. Harris, and R. Hledik, "Unlocking the 53 billion savings from smart meters in the EU: how increasing the adoption of dynamic tariffs could make or break the EU›s smart grid investment,"Energy Policy, vol. 38, no. 10, pp. 6222–6231, 2010. · ·

50. R. G. Pratt, P. Balducci, C. Gerkensmeyer et al., "TheSmartGrid: an estimation of the energy and CO_2benefits," Research Report, Revision 1 PNNL-19112, The U.S. Department of Energy by Pacific Northwest National Laboratory, 2010.

51. A. F. Alhajji, "What is energy security? definitions and concepts (part 3/5)," Middle East Economic Survey, vol. 50, no. 45, 2007.

52. J. M. Chevalier, "Security of energy supply for the European union," European Review of Energy Markets, vol. 1, no. 3, pp. 1–20, 2006.

53. IEA, World Energy Outlook 2007—China and India Insights, International Energy Agency, Paris, France, 2007.

54. APERC, "A Quest for Energy Security in the 21st century; Institute of energy economics," Japan, 2007.

55. C. Linde, M. P. Amineh, A. Corelje, D. Jong, and S. Hansen, "EU energy supply security and geopolitics (Tren/C1-06-2002) study," Tech. Rep., Clingendael International Energy Programme (CIEP), The Hague, The Netherlands, 2004, http://reaccess.epu.ntua.gr/LinkClick.aspx?fileticket=Bgw7mGJEWls%3D.

56. B. Kruyt, D. P. vanVuuren, H. J. M. deVries, and H. Groenenberg, "Indicators for energy security,"Energy Policy, vol. 37, no. 6, pp. 2166–2181, 2009. · ·

57. POSTNOTE, "Measuring Energy Security," 2012.

58. USGS, World Petroleum Assessment, 2000.

59. F. M. M. Mulders, J. M. M. Hettelar, and F. van Bergen, Assessment of the Global Fossil Fuel Reserves and Resources for TIMER, TNO Built Environment and Geosciences, Utrecht, The Netherlands, 2006.

60. A. F. Alhajji and L. W. James, "Measures of Petroleum Dependenceand Vulnerability in OECD Countries," Middle East Economic Survey, 2003.

61. J. C. Jansen, W. G. Arkel, and M. G. Boots, "Designing indicators of long-term energy supply security," ECN-C-04-007; 35, 2004.

62. IEA, Energy Security and CLimate Change Policy Interactions, An Assesment Framework, 2004.

63. Stirling, "On the economics and analysis of diversity," SPRU Electronic Working Papers Series, Paper 28, 1999.

64. J. C. Jansen, W. G. van Arkel, and M. G. Boots, "Designing indicators of longenergy supply security," 2004.

65. J. Martchamadol and S. Kumar, "Thailand›s energy security indicators," Renewable and Sustainable Energy Reviews, vol. 16, no. 8, pp. 6103–6122, 2012. · ·

66. IEA, "Energy Security and Climate Policy," 2007.

67. M. K. Datar, "Stock market liquidity: measurementand implications," in Proceedings of the 4th Capital Market Conference, 2000.

68. A. Molderink, V. Bakker, M. G. C. Bosman, J. L. Hurink, and G. J. M. Smit, "Domestic energy management methodology for optimizing efficiency in smart grids," in Proceedings of the IEEE Bucharest PowerTech, pp. 1–7, Bucharest, Romania, June-July 2009. · ·

69. L. M. Ayompe, A. Duffy, S. J. McCormack, and M. Conlon, "Validated real-time energy models for small-scale grid-connected PV-systems," Energy, vol. 35, no. 10, pp. 4086–4091, 2010. · ·

70. H. Lund, "Large-scale integration of wind power into different energy systems," Energy, vol. 30, no. 13, pp. 2402–2412, 2005. · ·

71. J. A. P. Lopes, N. Hatziargyriou, J. Mutale, P. Djapic, and N. Jenkins, "Integrating distributed generation into electric power systems: a review of drivers, challenges and opportunities," Electric Power Systems Research, vol. 77, no. 9, pp. 1189–1203, 2007. · ·

72. A. A. Bayod-Rújula, "Future development of the electricity systems with distributed generation," Energy, vol. 34, no. 3, pp. 377–383, 2009. · ·

73. M. Rönnelid, B. Perers, and B. Karlsson, "Construction and testing of a large-area CPC-collector and comparison with a flat plate collector," Solar Energy, vol. 57, no. 3, pp. 177–184, 1996. · ·

74. N. C. Batista, R. Melício, J. C. O. Matias, and J. P. S. Catalão, "Photovoltaic and wind energy systems monitoring and building/home energy management using ZigBee devices within a smart grid," Energy, vol. 49, no. 1, pp. 306–315, 2013. · ·

75. P. Vytelingum, T. D. Voice, S. D. Ramchurn, and A. Rogers, "Agent-based micro-storage management for the smart grid," in Proceedings of the 9th International Conference on Autonomous Agents and Multiagent Systems (AAMAS ‹10), vol. 1, pp. 39–46, May 2010.

76. A. Pina, C. Silva, and P. Ferrão, "The impact of demand side management strategies in the penetration of renewable electricity," Energy, vol. 41, no. 1, pp. 128–137, 2012. · ·

77. J. Oyarzabal, J. Jimeno, J. Ruela, A. Engler, and C. Hardt, "Agent based Micro Grid Management System," in Proceedings of the International Conference on Future Power Systems, pp. 6–11, IEEE, Amsterdam, Netherlands, November 2005. · ·

78. S. Abu-Sharkh, R. J. Arnold, J. Kohler et al., "Can microgrids make a major contribution to UK energy supply?" Renewable and Sustainable Energy Reviews, vol. 10, no. 2, pp. 78–127, 2006. · ·

79. J. McDonald, "Adaptive intelligent power systems: active distribution networks," Energy Policy, vol. 36, no. 12, pp. 4346–4351, 2008. · ·

80. V. H. Mendez, J. Rivier, J. I. Fuente, T. Gomez, and J. Arceluz, Impact of Distribute Generation on Distribution Network, Universidad Pontificia Comillas, Madrid, Spain, 2002.

81. J. K. Kaldellis, "Integrated electrification solution for autonomous electrical networks on the basis of RES and energy storage configurations," Energy Conversion and Management, vol. 49, no. 12, pp. 3708–3720, 2008. · ·

82. A. Sheikhi, A. M. Ranjbar, and M. Oraee, "Optimal operation and size for an energy hub with CCHP,"Energy and Power Engineering, vol. 3, pp. 641–649, 2011.

83. K. N. Finney, Q. Chen, V. N. Sharifi et al., "Developments to an existing city-wide district energy network: part II—analysis of environmental and economic impacts," Energy Conversion and Management, vol. 62, pp. 176–184, 2012. · ·

84. T. M. I. Mahlia and P. A. A. Yanti, "Cost efficiency analysis and emission reduction by implementation of energy efficiency standards for electric motors," Journal of Cleaner Production, vol. 18, no. 4, pp. 365–374, 2010. · ·

85. IEA, "Renewable energy outlook," Global Energy Trends, 2013.

86. "Smart Grid Apps: Six trends that will shape grid evolution," 2011.

87. R. G. Pratt, P. J. Balducci, T. F. Sanquist et al., The Smart Grid: An Estimation of the Energy and CO2 Benefits, R. Pacific Northwest National Laboratory, Washington, DC, USA, 2010.

88. NEB, Suruhanjaya Tenaga (Energy Commission), National Energy Balance, Pusat Tenaga, Malaysia, 2012.

89. REN21, "Renewables 2012," Global Status Report, 2012.

90. TRSG, "Technology Roadmap Smart Grids," International Energy Agency 2011, 2011,http://www.iea.org/publications/freepublications/publication/smartgrids_roadmap.pdf.

91. NIST, IST Special Publication 1108, NIST Framework and Roadmap for Smart Grid Interoperability Standards, Release 1.0, January 2010,http://www.nist.gov/public_affairs/releases/upload/smartgrid_interoperability_final.pdf.

92. J. J. Venkatesh, "Renewable Energy integration in Smart girds," Smart Grid Seminar, 2014,http://cseweb.ucsd.edu/~trosing/lectures/cse291_renew_store.pdf.

93. PROGRASS, 2014, http://www.prograss.eu/index.php?id=75.

94. "Game to be green Grothermal power," 2014.

Swarm Intelligence-Based Smart Energy Allocation Strategy for Charging Stations of Plug-in Hybrid Electric Vehicles

Imran Rahman[1], Pandian M. Vasant[1], Balbir Singh Mahinder Singh[1], and M. Abdullah-Al-Wadud[2]

[1]Department of Fundamental and Applied Sciences, Universiti Teknologi PETRONAS, Bandar Seri Iskandar, 31750 Tronoh, Perak, Malaysia

[2]Department of Software Engineering, College of Computer and Information Sciences, King Saud University, Saudi Arabia

ABSTRACT

Recent researches towards the use of green technologies to reduce pollution and higher penetration of renewable energy sources in the transportation sector have been gaining popularity. In this wake, extensive participation of plug-in hybrid electric vehicles (PHEVs) requires adequate charging allocation strategy using a combination of smart grid systems and smart charging infrastructures. Daytime charging stations will be needed for daily usage of PHEVs due to the limited all-electric range. Intelligent energy management is an important issue which has already drawn much attention of researchers. Most of these works require formulation of mathematical models with extensive use of computational intelligence-based optimization techniques to solve many technical problems. In this paper, gravitational search algorithm (GSA) has been applied and compared with another member of swarm family, particle swarm optimization (PSO), considering constraints such as energy price, remaining battery capacity, and remaining charging time. Simulation results obtained for maximizing the highly nonlinear objective function evaluate the performance of both techniques in terms of best fitness.

INTRODUCTION

The vehicular network recently accounts for around 25% of CO_2 emissions and over 55% of oil consumption around the world [1]. Carbon dioxide (CO_2) is the primary greenhouse gas emitted through human activities like combustion of fossil fuels (coal, natural gas, and oil) for energy and transportation. Several researchers have proved that a great amount of reductions in greenhouse gas emissions and the increasing dependence on oil could be accomplished by electrification of transport sector [2]. Indeed, the adoption of hybrid electric vehicles (HEVs) has brought significant market success over the past decade. Vehicles can be classified into three groups: internal combustion engine vehicles (ICEVs), hybrid electric vehicles (HEVs), and all-electric vehicles (AEVs) [3]. Plug-in hybrid electric vehicles (PHEVs) which are very recently introduced promise to boost up the overall fuel efficiency by holding a higher capacity battery system, which can be directly charged from traditional power grid system that helps the vehicles to operate continuously in "all-electric range" (AER). All-electric vehicle or AEV is a vehicle using electric power as the only source to move the vehicle [4]. Plug-in hybrid electric vehicles with a connection to the smart grid can possess all of these strategies. Hence, the widely extended adoption of PHEVs might play a significant role in the alternative energy integration into traditional grid systems [5]. There is a need of efficient mechanisms and algorithms for smart grid technologies in order to solve highly heterogeneous problems like energy management, cost reduction, efficient charging infrastructure, and so forth with different objectives and system constraints [6].

According to a statistics of Electric Power Research Institute (EPRI), about 62% of the entire United States (US) vehicle will comprise PHEVs within the year 2050 [7]. Moreover, there is an increasing demand to implement this technology on the electric grid system. Large numbers of PHEVs have the capability to threaten the stability of the power system. For example, in order to avoid interruption when several thousand PHEVs are introduced into the system over a short period of time, the load on the power grid will need to be managed very carefully. One of the main targets is to facilitate the proper interaction between the power grid and the PHEV. For the maximization of customer satisfaction and minimization of burdens on the grid, a complicated control mechanism will need to be addressed in order to govern multiple battery loads from a number of PHEVs appropriately [8]. The total demand pattern will also have an important impact on the electricity industry due to differences in the needs of the PHEVs parked in the deck at certain time [9]. Proper management can ensure strain minimization of the grid and enhance the transmission and generation of electric power supply. The control of PHEV charging depending on the locations can be classified into two groups: household charging and public charging. The proposed optimization focuses on the public charging station for plug-in vehicles because most of PHEV charging is expected to take place in public charging locations [10].

Wide penetration of PHEVs in the market depends on a well efficient charging infrastructure. The power demand from this new load will put extra stress on the traditional power grid [11]. As a result, a good number of PHEV charging infrastructures with appropriate facilities are essential to be built for recharging electric vehicles; for this some strategies have been proposed by the researchers [12, 13]. Charging stations are needed to be built at workplaces, markets/shopping malls, and home. In [14], authors proposed the necessity of building new smart charging station with effective communication among utilities along with substation control infrastructure in view of grid stability and proper energy utilization. Furthermore, assortment of charging stations with respect to charging characteristics of different PHEVs traffic mobility characteristics, sizeable energy storage, cost minimization, quality of services (QoS), and optimal power of intelligent charging station are underway [15]. Thus, evolution of reliable, efficient, robust, and economical charging infrastructure is underway. In this wake, numerous techniques and methods have been proposed for deployment of charging station for PHEVs [16, 17].

One of the important constraints for accurate charging is state of charge (SoC). Charging algorithm can accurately be managed by the precise state of charge estimation [18]. An approximate graph of a typical lithium-ion cell voltage versus SoC is shown in Figure 1. The figure indicates that the slope of the curve below 20% and above 90% is high enough to result in a detectable voltage difference to be relied on by charge balancing control and measurement circuits [19]. There is a need of in-depth study on maximization of average SoC in order to facilitate intelligent energy allocation for PHEVs in a charging station. Gravitational search algorithm (GSA) is one of the newest heuristic algorithms introduced by Rashedi et al. [20]. GSA algorithm is also a member of swarm intelligence family which is inspired by the well-known law of gravity and interactions between the masses and implements Newtonian gravity and the laws of motion [21–23].

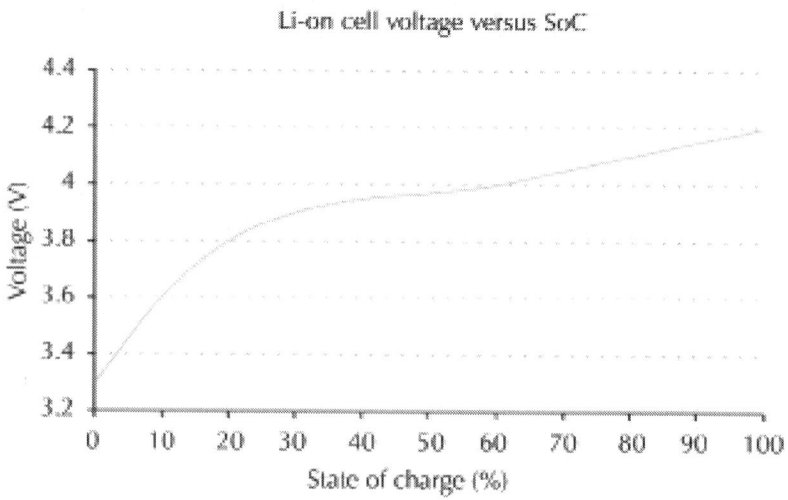

Figure 1: Li-ion cell voltage versus state of charge [24].

GSA-based optimization has already been used by the researchers for postoutage bus voltage magnitude calculations, economic dispatch with valve-point effects, optimal sizing and suitable placement for distributed generation (DG) in distribution system, optimization of synthesis gas production [25], rectangular patch antenna [26], orthogonal array based performance improvement [27], solving thermal unit commitment (UC) problem, and finding out optimal solution for optimal power flow (OPF) problem in a power system [28]. Specifically, we are investigating the use of the gravitational search algorithm (GSA) method for developing real-time and large-scale optimizations for allocating power.

The remainder of this paper is organized as follows: next section will describe the specific problem that we are trying to solve. We will provide the optimization objective and constraints and mathematical formulation of our algorithm, review the GSA method, and describe how the algorithm works for our optimization problems. The simulation results and analysis are then presented. Finally, conclusions and future directions are drawn.

PROBLEM FORMULATION

The idea behind smart charging is to charge the vehicle when it is most beneficial such as when the electricity price and total power demand remain lowest or there is excess capacity of generated power [24].

Suppose there is a charging station with the capacity of total power P.N Total numbers of PHEVs need to be charged within 24 hours of time interval. The proposed system should allow PHEVs to leave the charging station before their expected leaving time for making the system more effective. It is worth to mention that each PHEV is regarded to be plugged in to the charging station once. The main aim is to allocate power intelligently for each PHEV coming to the charging station. The state of charge is the main parameter which needs to be maximized in order to allocate power effectively. For this, the objective function considered in this paper is the maximization of average SoC and thus allocates energy for PHEVs at the next time step. The constraints considered are charging time, present SoC, and price of the energy.

The objective function is defined as

$$\max J(k) = \sum_i w_i(k) \, SoC_i(k+1),$$

$$w_i(k) = f\left(C_{r,i}(k), T_{r,i}(k), D_i(k)\right),$$

$$C_{r,i}(k) = \left(1 - SoC_i(k)\right) * C_i,$$

$$(1)$$

where $C_{r,i}(k)$ is the battery capacity (remaining) needed to be filled for i number of PHEV at time step k; C_i is the battery capacity (rated) of the i number of PHEV; remaining time for charging a particular PHEV at time step k is expressed as $T_{r,i}(k)$; the price difference between the real-time energy price and the price that a specific customer at the i number of PHEV charger is willing to pay at time step k is presented by $D_i(k)$; $w_i(k)$ is the charging weighting term of the number of PHEV at time step (a function of charging time, present SoC, and price of the energy); $SoC_i(k+1)$ is the state of charge of the number of PHEV at time step .

Here, the weighting term indicates a bonus proportional to the attributes of a specific PHEV. For example, if a PHEV has a lower initial SoC and less charging time (remaining), but the driver is eager to pay a higher price, the system will provide more power to this particular PHEV battery charger:

$$w_i(k) \, \alpha \left[Cap_{r,i}(k) + D_i(k) + \frac{1}{T_{r,i}}(k) \right].$$

$$(2)$$

The charging current is also assumed to be constant over Δt :

$$\left[SoC_i(k+1) - SoC_i(k)\right] \cdot Cap_i = Q_i = I_i(k)\,\Delta t,$$

$$SoC_i(k+1) = SoC_i(k) + \frac{I_i(k)\,\Delta t}{Cap_i},$$

$$(3)$$

where the sample Δt time is defined by the charging station operators and $I_i(k)$ is the charging current over Δt .

The battery model is regarded as a capacitor circuit, where C_i is the capacitance of battery (Farad). The model is defined as

$$C_i \cdot \frac{dV_i}{dt} = I_i.$$

(4)

Therefore, over a small time interval, one can assume the change of voltage to be linear:

$$C_i \cdot \frac{[V_i(k+1) - V_i(k)]}{\Delta t} = I_i,$$

$$V_i(k+1) - V_i(k) = \frac{I_i \Delta t}{C_i}.$$

(5)

As the decision variable used here is the allocated power to the PHEVs, by replacing $I_i(k)$ with $P_i(k)$ the objective function finally becomes

$$J(k) = \sum w_i \cdot \left[SOC_i(k) + (2P_i(k)\Delta t) \right.$$

$$\times \left(0.5 \cdot C_i \cdot \left[\sqrt{\frac{2P_i(k)\Delta t}{C_i} + V_i^2(k)} \right. \right.$$

$$\left. \left. \left. + V_i(k) \right] \right)^{-1} \right].$$

(6)

System Constraints

Possible real-world constraints could include the charging rate (i.e., slow, medium, and fast), the time that the PHEV is connected to the grid, the desired departure SOC, the maximum electricity price that a user is willing to pay, and certain battery requirements. Furthermore, the available communication bandwidth could limit sampling time, which would have effects on the processing ability of the vehicle.

Power obtained from the utility ($P_{utility}$) and the maximum power ($P_{i,max}$) absorbed by a specific PHEV are the primary energy constraints being considered in this paper. The power demand of a PHEV/PEV cannot exceed the rated power output of the battery charger:

$$\sum_i P_i(k) \leq P_{utility}(k) \times \eta,$$

$$0 \leq P_i(k) \leq P_{i,max}(k).$$

(7)

The overall charging efficiency of a particular charging infrastructure is described by η. From the system point of view, charging efficiency is supposed to be constant at any given time step. Maximum battery SoC limit for i the number of PHEV is $SoC_{i,max}$. When SoC_i reaches the values close to $SoC_{i,max}$, the i number of battery charger shifts to a standby mode. The state of charge ramp rate is confined within limits by the constraint ΔSoC_{max}. To accommodate the system dynamics, the energy scheduling is updated when when (i) system utility data is updated; (ii) a new PHEV is plugged in, and (iii) time period t has periodically passed. Table 1 shows all the objective function parameters that were tuned for performing the optimization.

Table 1: Parameter settings of the objective function

Parameter	Values
Fixed parameters	Maximum power, $P_{i,max}$ = 6.7 kWh
	Charging station efficiency, η = 0.9
	Total charging time, Δt = 20 minutes (1200 seconds)
	Power allocation to each PHEV: 30 W
Variables	0.2 ≤ state of charge (SoC) ≤ 0.8
	Waiting time ≤ 30 minutes (1800 seconds)
	16 kWh ≤ battery capacity (C_i) ≤ 40 kWh
Constraints	$\sum_i P_i(k) \le P_{utility}(k) \times \eta$
	$0 \le P_i(k) \le P_{i,max}(k)$ $0 \le SoC_i(k) \le SoC_{i,max}$
	$0 \le SoC_i(k+1) - SoC_i(k) \le \Delta SoC_{max}$

Energy allocation to PHEV charging station is subjected to various constraints as mentioned in the problem formulation section. Different constraints make the entire search space limited to a particular suitable region. So, a powerful optimization algorithm should be implemented in order to achieve high quality solutions with a stable convergence rate.

GRAVITATIONAL SEARCH ALGORITHM

GSA is an optimization method which has been introduced by Rashedi et al. in the year of 2009 [20]. In GSA, the specifications of each mass (or agent) are four in total, which are inertial mass, position, active gravitational mass and passive gravitational mass. The position of the mass presents a solution of a particular problem and masses (gravitational and inertial) are obtained by using a fitness function. GSA can be considered as a collection of agents (candidate solutions), whose masses are proportional to their value of fitness function. During generations, all masses attract each other by the gravity forces between them. A heavier mass has the bigger attraction force. Therefore the heavier masses which are probably close to the global optimum attract the other masses proportional to their distances.

Law of Gravity

The law states that particles attract each other and the force of gravitation between two particles is directly proportional to the product of their masses and inversely proportional to the distance between them.

Law of Motion

The law states that the present velocity of any mass is the summation of the fraction of its previous velocity and the velocity variance. Variation in the velocity or acceleration of any mass is equal to the force divided by inertia mass.

The gravitational force is expressed as follows:

$$F_{ij}^d(t) = G(t) \frac{M_{pi}(t) \times M_{aj}(t)}{R_{ij}(t) + \varepsilon} \left(x_j^d(t) - x_i^d(t) \right),$$

$$(8)$$

where $M_{a,j}$ is the active gravitational mass related to agent j, M_{pi} is the passive gravitational mass related to agent i, $G(t)$ is gravitational constant at time t, ε is a small constant, and $R_{ij}(t)$ is the Euclidian distance between two agents i and j. The $G(t)$ is calculated as follows:

$$G\left(t\right) = G_0 \times \exp\left(\frac{-\alpha \times \text{iter}}{\text{max iter}}\right),$$

(9)

where   and G_0 are descending coefficient and primary value, respectively, and current iteration and maximum number of iterations are expressed as iter and max iter. In a problem space with the dimension d, the overall force acting on agent i is estimated as the following equation:

$$F_i^d\left(t\right) = \sum_{j=1, j \neq i}^{N} \text{rand}_j F_{ij}^d\left(t\right),$$

(10)

where rand_j is a random number with interval [0,1]. From law of motion we know that an agent's acceleration is directly proportional to the resultant force and inverse of its mass, so the acceleration of all agents should be calculated as follows:

$$\text{ac}_i^d\left(t\right) = \frac{F_i^d\left(t\right)}{M_{ii}\left(t\right)},$$

(11)

where t is a specific time and M_{ii} is the mass of the object i. The velocity and position of agents are calculated as follows:

$$\text{vel}_i^d\left(t + 1\right) = \text{rand}_i \times \text{vel}_i^d\left(t\right) + \text{ac}_i^d\left(t\right),$$

(12)

$$x_i^d\left(t + 1\right) = x_i^d\left(t\right) + \text{vel}_i^d\left(t + 1\right),$$

(13)

where rand_i is a random number with interval [0,1].

Gravitational and inertia masses are simply calculated by the fitness evaluation. A heavier mass means a more efficient agent. This means that better agents have higher attractions and walk more slowly. Assuming the equality of the gravitational and inertia mass, the values of masses are calculated using the map of fitness. We update the gravitational and inertial masses by the following equations:

$$M_{ai} = M_{pi} = M_{ii} = M_i, \quad i = 1, 2, \ldots, N.$$

(14)

In gravitational search algorithm, all agents are initialized first with random values. Each of the agents is a candidate solution. After initialization, velocities for all agents are defined using (12). Moreover, the gravitational constant, overall forces, and accelerations are determined by (9), (10), and (11), respectively. The positions of agents are calculated using (13). At the end, GSA will be terminated by meeting the stopping criterion of maximum 100 iterations. The parameter settings for GSA are demonstrated in Table 2. Moreover, GSA flowchart is shown in Figure 2.

Table 2: GSA parameter settings

Parameters	Values
Primary parameter, G_0	100
Number of mass agents, n	100
Constant parameter, α	20
Constant parameter, ε	.01
Power of "R"	1
Maximum iteration	100
Number of runs	50

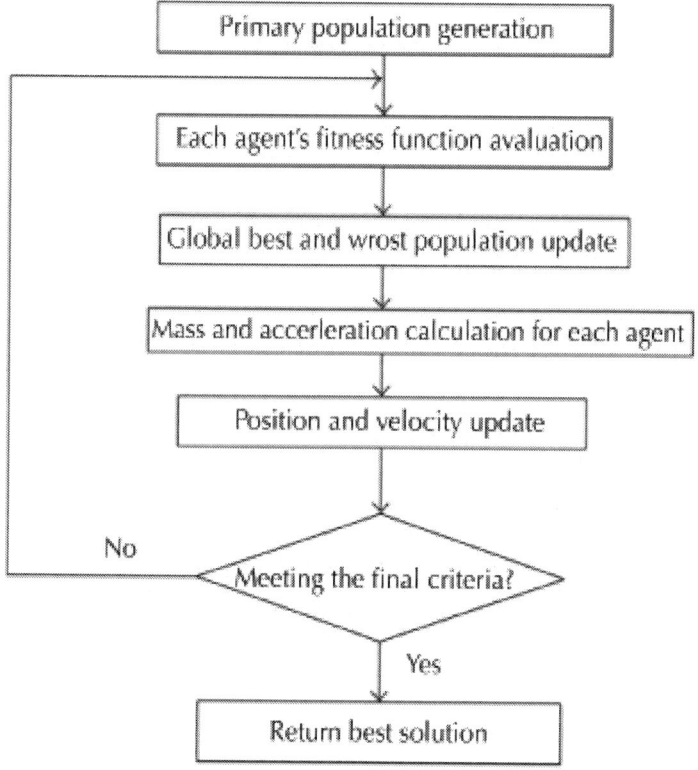

Figure 2: The GSA flowchart.

SIMULATION RESULTS AND ANALYSIS

The GSA algorithm was applied to find out global best fitness of the objective function (Algorithm 1). All the calculations were run on an Intel (R) Core i5-3470 M CPU@ 3.20 GHz, 4.00 GB RAM, Microsoft 32 bit Windows 7 OS, and MATLAB© R2013a.

Algorithm 1

(1) Initialization of total N mass agents randomly
(2) Computation of G(t), Fitness (Best and Worst)
(3) For each of the agent I, evaluate:
(3.1) Fitness$_i$
(3.2) Mass$_i$
(3.3) Force of Mass$_i$
(3.4) Acceleration of Mass$_i$
(3.5) Mass$_i$ velocity update
(3.6) New position of Agent$_i$
If (Probability$_i$>Thershold)
{
If (Probability$_i$>Rand$_j$)
Then return Best Fitness
solution so far
Else
Modification of solution
}
(4) Failed to meet stopping criteria,
Go To Step 2, Else Stop

Many optimization algorithms involve local search techniques which can get stuck on local maxima. Most search techniques strive to find a global maximum in the presence of local maxima [29]. One of the most important characteristics of GSA is its significant performance during exploration process. The capability of an algorithm to extend the problem in search gap is known as exploration while the ability of an algorithm to recognize optimal solution near a favorable one is exploitation [30, 31].

Figures 3, 4, 5, 6, and 7 show the simulation results for 50,100, 300,500, and 1000 plug-in hybrid electric vehicles (PHEVs), respectively, for finding the maximum fitness value of objective function J. In order to evaluate the performance and show the efficiency and superiority of the proposed algorithm, we ran each scenario 50 times in total.

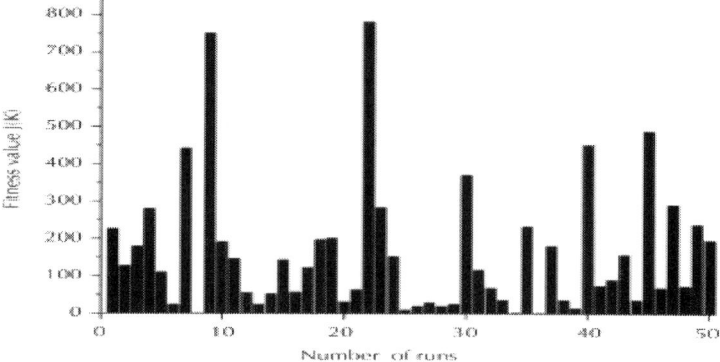

Figure 3: Fitness value versus number of runs (50 PHEVs).

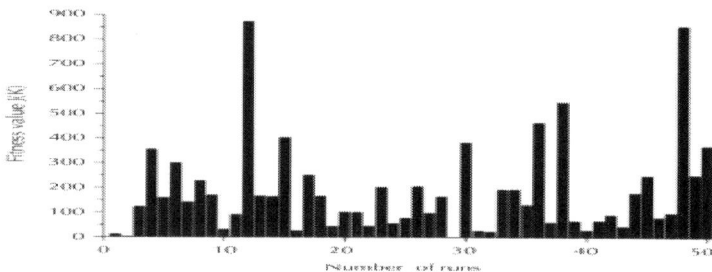

Figure 4: Fitness value versus number of runs (100 PHEVs).

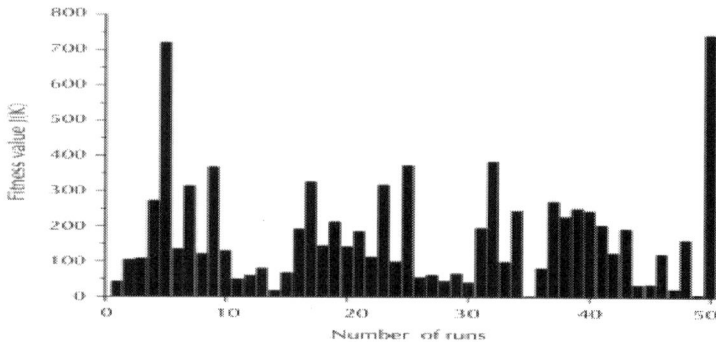

Figure 5: Fitness value versus number of runs (300 PHEVs).

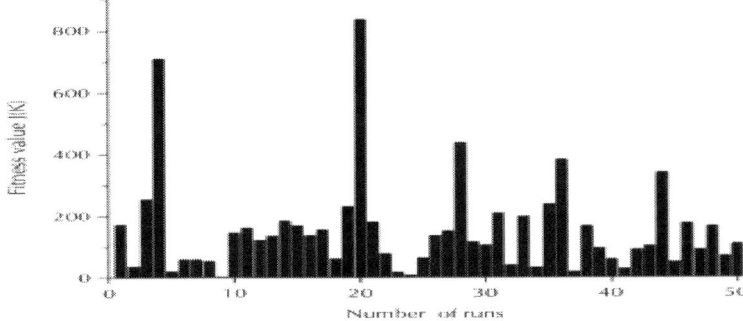

Figure 6: Fitness value versus number of runs (500 PHEVs).

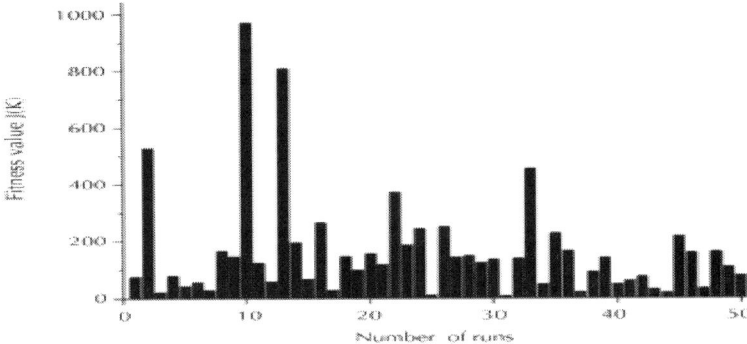

Figure 7: Fitness value versus number of runs (for 1000 PHEVs).

For Figure 3 (50 PHEVs), the maximum best fitness and minimum best fitness were 781.1267 and 0.2191, respectively.

The average best fitness is 158.8289. Figure 4 depicts the maximum fitness value for 100 PHEVs. In this case, the maximum best fitness and minimum best fitness were 579.3955 and 3.2523. The average best fitness is decreased into 139.7536.

For Figure 5 (300 PHEVs), the maximum best fitness and minimum best fitness were 743.1251 and 2.3279, respectively. The average best fitness is 172.4296.

Figure 6 depicts the maximum fitness value for 500 PHEVs. In this case, the maximum best fitness and minimum best fitness were

836.2707 and 0.9818. The average best fitness is decreased into 152.36437.

Figure 7 shows the maximum fitness value for 1000 PHEVs. In this case, the maximum best fitness and minimum best fitness were 968.7652 and 7.2747. The average best fitness is decreased into 161.52349.

Finally, Table 3 summarizes the result. From that it can be concluded that average best fitness remains almost in similar pattern for five (05) different scenarios.

Table 3: Fitness evaluation of GSA

Fitness function J(k)	For 50 PHEVs	For 100 PHEVs	For 300 PHEVs	For 500 PHEVs	For 1000 PHEVs
Maximum best fitness	781.1267	579.3955	743.1251	836.2707	968.7652
Average best fitness	158.8289	139.7536	172.4296	152.36437	161.52349
Minimum best fitness	0.2191	3.2523	2.3279	0.9818	7.2747

Performance Evaluation of GSA

Convergence Analysis

It can be apparently seen that although the algorithm has been set to run for maximum 100 iterations, the convergence happened in about 20 iterations. The result derived in this paper reveals that each object of the standard GSA converges to a stable point. Here, the assumption was that the gravitational and inertia masses are the same. However, for some applications different values for them can be used. A heavier inertia mass provides a slower motion of agents in the search space and hence a more precise search [20]. On the contrary, a heavier gravitational mass causes a higher attraction of agents. This allows a faster convergence. The analysis results confirm the convergence characteristics of GSA according to the given parameters ranges of the algorithm. Figures 8, 9, 10, 11, and 12 show the convergence behavior of GSA. The best fitness function shows convergences after the same iterations (35 iterations) for both 50 and 100 numbers of PHEVs while for 500 and 1000 numbers of PHEVs, it shows early convergence (before 20 iterations).

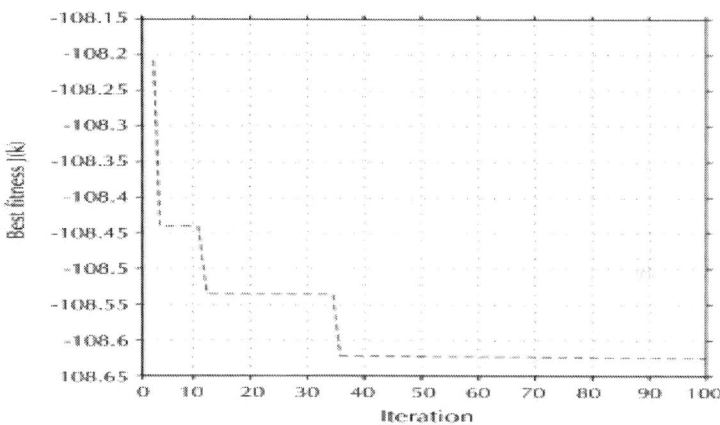

Figure 8: Best fitness versus iteration (50 PHEVs).

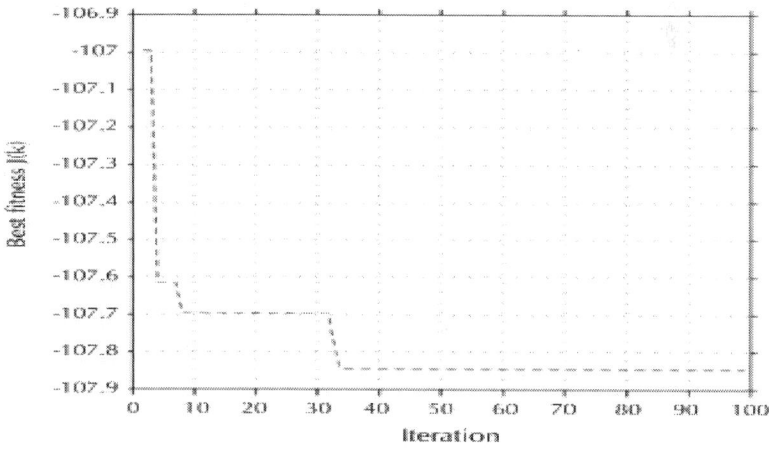

Figure 9: Best fitness versus iteration (100 PHEVs).

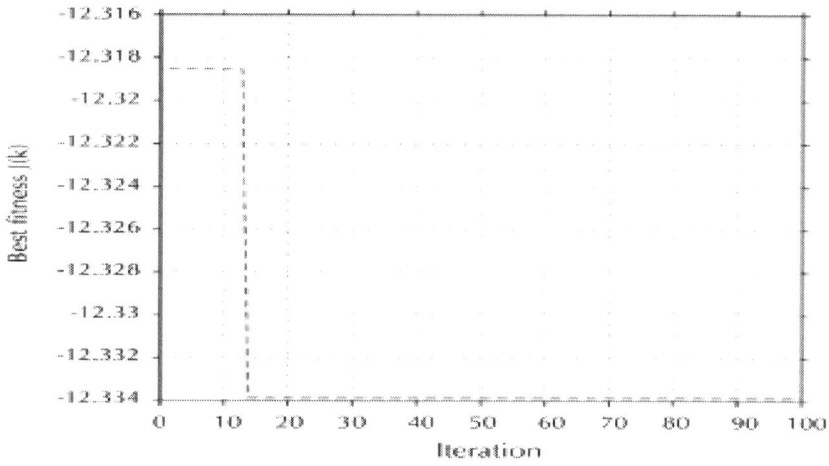

Figure 10: Best fitness versus iteration (300 PHEVs).

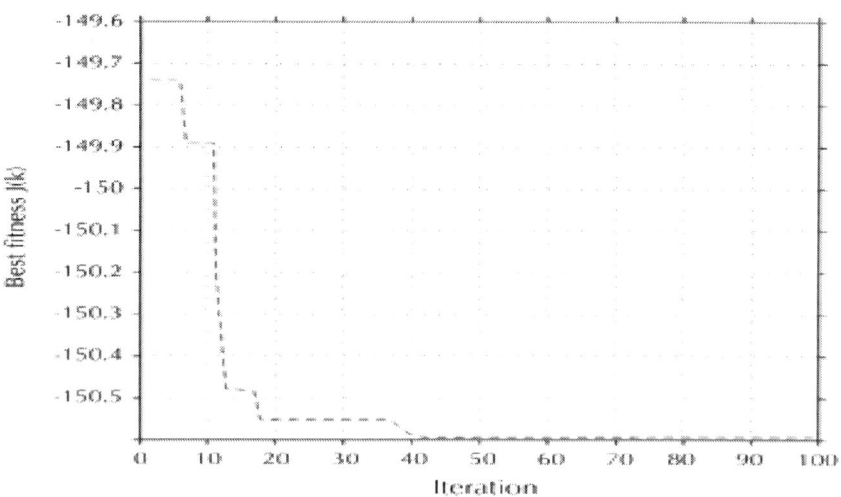

Figure 11: Best fitness versus iteration (500 PHEVs).

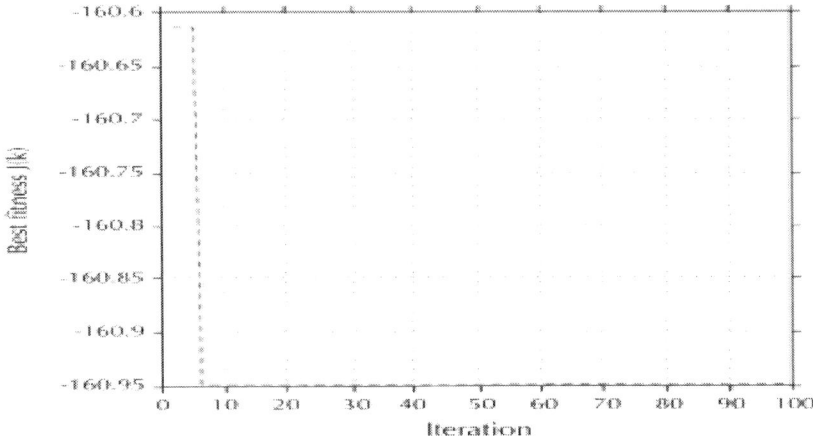

Figure 12: Best fitness versus iteration (1000 PHEVs).

Robustness

The similar numeric patterns of average best fitness show the robustness of GSA method. This method resists change without adapting its initial stable configuration for different cases (number of PHEVs) which proves GSA as a robust algorithm.

Diversity

Here, the average best fitness gives different values with the increment of PHEVs population. The rate of convergence of mass agents in GSA is good through the fast information flowing among mass agents, so its diversity decreases very quickly in the successive iterations and leads to a suboptimal solution.

Computational Cost

Here, we measured the computational cost of the algorithm in terms of total running time. Table 6 shows the computational time of GSA for five different scenarios. Here, average CPU time is measured in seconds. As GSA needs a good number of parameter tuning, the computational cost increases with the increment of the total number of PHEVs.

Quality of Solution

When an algorithm finds an optimal solution to a given problem, one of the important factors is speed and rate of convergence to the optimal solution. For heuristics, the additional consideration of how close the heuristic solution comes to optimal is generally the primary concern of the researcher [32]. In GSA, the faster convergence and better exploitation rate ensure good quality solution, which is best fitness function.

For this optimization, the initial state of charge was expressed as a random number which is continuous and uniform between 0.2 and 0.6. The sample time was set around 1200 seconds (20 minutes). The remaining charge time was defined as continuous random number between 0 and 6 hours. The price according to customer's choice for paying the bill for electricity was expressed as a continuous random number which is in between $1 and $2.

The capacity of the battery was assumed to be identical for all vehicles.

Comparison between GSA and PSO

Particle swarm optimization (PSO) with the parameter settings stated in Table 4 was also performed for the same objective function and compared with the performance of gravitational search algorithm in terms of average best fitness. The swarm size and maximum iterations were set exactly the same as those of GSA algorithm for the comparison purpose. The values of parameters c1, c2, and w were set as standard values, 1.4, 1.4, and 0.9, respectively.

Table 4: PSO parameter settings

Parameters	Values
Size of the swarm	100
Maximum number of steps	100
PSO parameter,C_1	1.4
PSO parameter,C_2	1.4

PSO inertia (w)	0.9
Maximum iteration	100
Number of runs	50

Table 5: Summarizes the comparisons of GSA with PSO algorithm in terms of average best fitness

Average best fitness for	PSO	GSA
50 PHEVs	142.839	158.8289
100 PHEVs	171.102	182.3097
300 PHEVs	169.312	172.4296
500 PHEVs	150.869	152.36437
1000 PHEVs	156.802	161.52349

Table 6: Computational time for PSO and GSA

Computational Time (sec.)	PSO	GSA
50 PHEVs	1.650	2.721
100 PHEVs	1.686	4.439
500 PHEVs	1.990	18.165
1000 PHEVs	2.398	36.275

From Figure 13 it is clear that gravitational search algorithm outperformed particle swarm optimization in terms of average best fitness. Starting from 50 numbers of PHEVs up to 1000 PHEVs, GSA shows better fitness value than PSO.

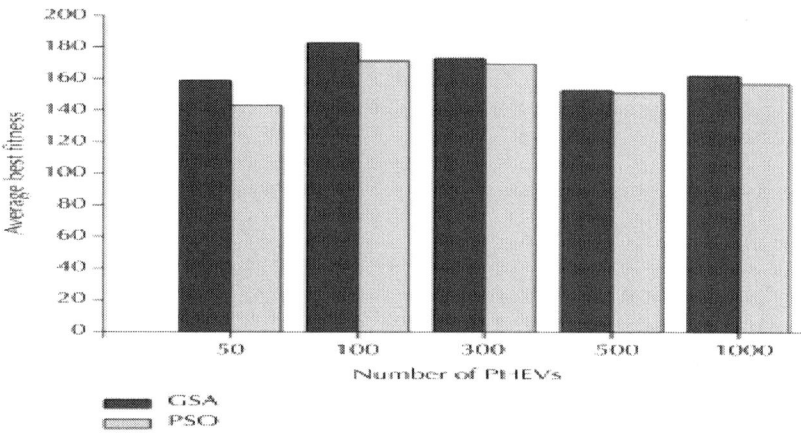

Figure 13: Average best fitness versus number of PHEVs.

Table 7 illustrates the advantages and disadvantages of both GSA and PSO for solving different optimization problems.

Table 7: Advantages and disadvantages of PSO and GSA

Optimization method	Advantages	Disadvantages
PSO	Less parameters tuning Easy constraint Good for multiobjective optimization [33]	Low quality solution Needs memory to update velocity Slow convergence rate
GSA	High quality solution Good convergence rate Local exploitation capability [34]	Needs more computational time More parameters tuning

It has been proven that gravitational search algorithm has good ability to search for the global optimum, but it suffers from slow searching speed in the last iterations [35]. Moreover, the inertia mass is against the motion and slows the mass movement. Agents with heavy inertia mass move slowly and hence search the space more locally. So, it can be considered as an adaptive learning rate [34]. GSA is a memory-less algorithm. However, it works competently like the algorithms with memory. Our simulation results show the good convergence rate of the GSA.

CONCLUSION AND RECOMMENDATIONS

In this paper, gravitational search algorithm- (GSA-) based optimization was performed in order to optimally allocate power to each of the PHEVs entering into the charging station. A sophisticated controller will need to be designed in order to allocate power to PHEVs appropriately. For this wake, the applied algorithm in this paper is a step towards real-life implementation of such controller for PHEV charging infrastructures.

Here, five (05) different numbers of PHEVs were considered for MATLAB© simulation and then obtained results were compared with PSO in terms of average best fitness. The success of the electrification of transportation sector solely depends on charging infrastructure. Only proper charging control and infrastructure management can assure the larger penetration of PHEVs. The researchers should try to develop efficient control mechanism for charging infrastructure in order to facilitate upcoming PHEVs penetration in highways. In future, more vehicles should be considered for intelligent power allocation strategy and improved versions of GSA and hybrid swarm intelligence based methods should be applied to ensure low computational cost.

ACKNOWLEDGMENTS

The authors would like to thank Universiti Teknologi Petronas for supporting the research under Graduate Assistance Scheme. This research paper is financially sponsored by the Centre of Graduate study with the support of the Department of Fundamental & Applied Sciences, Universiti Teknologi Petronas.

REFERENCES

1. On certain integrals International Energy Agency (IEA), Transport, Energy and C02-Moving Towards Sustainability, 2009.

2. D. Holtz-Eakin and T. M. Selden, "Stoking the fires? CO_2 emissions and economic growth," Journal of Public Economics, vol. 57, no. 1, pp. 85–101, 1995.

3. S. F. Tie and C. W. Tan, "A review of energy sources and energy management system in electric vehicles," Renewable and Sustainable Energy Reviews, vol. 20, pp. 82–102, 2013.

4. "Environmental assessment of plug-in hybrid electric vehicles, Volume 1: Nationwide greenhouse gas emissions," Tech. Rep. 1015325, Electric Power Research Institute (EPRI), Palo Alto, Calif, USA, 2007.

5. H. Lund and W. Kempton, "Integration of renewable energy into the transport and electricity sectors through V2G," Energy Policy, vol. 36, no. 9, pp. 3578–3587, 2008.

6. A. R. Hota, M. Juvvanapudi, and P. Bajpai, "Issues and solution approaches in PHEV integration to the smart grid," Renewable and Sustainable Energy Reviews, vol. 30, pp. 217–229, 2014.

7. J. Soares, T. Sousa, H. Morais, Z. Vale, B. Canizes, and A. Silva, "Application-Specific Modified Particle Swarm Optimization for energy resource scheduling considering vehicle-to-grid," Applied Soft Computing Journal, 2013.

8. W. Su and M.-Y. Chow, "Computational intelligence-based energy management for a large-scale PHEV/PEV enabled municipal parking deck," Applied Energy, vol. 96, pp. 171–182, 2012.

9. W. Su and M. Y. Chow, "Performance evaluation of a PHEV parking station using particle swarm optimization," in Proceedings of the IEEE PES General Meeting, pp. 1–6, July 2011.

10. W. Su and M.-Y. Chow, "Performance evaluation of an EDA-based large-scale plug-in hybrid electric vehicle charging algorithm," IEEE Transactions on Smart Grid, vol. 3, no. 1, pp. 308–315, 2012.

11. K. Morrow, D. Karner, and J. Francfort, "Plug-in hybrid electric vehicle charging infrastructure review," Tech. Rep., The Idaho

National Laboratory, 2008.

12. "Investigation into the scope for the transport sector to switch to electric vehicles and plug- in hybrid vehicles," Tech. Rep., Department for Transport, London, UK, 2008.

13. D. Mayfield, "Site Design for Electric Vehicle Charging Stations, ver. 1.0," Sustainable Transportation Strategies, July 2012.

14. G. Boyle, Renewable Electricity and the Grid: The Challenge of Variability, Earthscan Publications, London, UK, 2007.

15. A. Hess, F. Malandrino, M. B. Reinhardt, C. Casetti, K. A. Hummel, and J. M. Barceló-Ordinas, "Optimal deployment of charging stations for electric vehicular networks," in Proceedings of the 1st ACM Conference on Urban Networking (UrbaNe '12), pp. 1–6, New York, NY, USA, December 2012.

16. Z. Li, Z. Sahinoglu, Z. Tao, and K. H. Teo, "Electric vehicles network with nomadic portable charging stations," in Proceeding of the IEEE 72nd Vehicular Technology Conference Fall (VTC '10), pp. 1–5, Ottawa, Canada, September 2010.

17. T. Ghanbarzadeh, S. Goleijani, and M. P. Moghaddam, "Reliability constrained unit commitment with electric vehicle to grid using hybrid particle swarm optimization and ant colony optimization," inProceeding of the IEEE PES General Meeting: The Electrification of Transportation and the Grid of the Future, pp. 1–7, San Diego, Calif, USA, July 2011.

18. A. Shafiei and S. S. Williamson, "Plug-in hybrid electric vehicle charging: current issues and future challenges," in Proceedings of the IEEE Vehicle Power and Propulsion Conference (VPPC '10), pp. 1–8, September 2010.

19. J. Yan, C. Li, G. Xu, and Y. Xu, "A novel on-line self-learning state-of-charge estimation of battery management system for hybrid electric vehicle," in Proceedings of the IEEE Intelligent Vehicles Symposium, pp. 1161–1166, June 2009.

20. E. Rashedi, H. Nezamabadi-pour, and S. Saryazdi, "GSA: a gravitational search algorithm," Information Sciences, vol. 179, no. 13, pp. 2232–2248, 2009.

21. D. Martens, B. Baesens, and T. Fawcett, "Editorial survey: swarm intelligence for data mining," Machine Learning, vol. 82, no. 1, pp. 1–42, 2011.

22.   J. Krause, J. Cordeiro, R. S. Parpinelli, and H. S. Lopes, "A survey of swarm algorithms applied to discrete optimization problems," in Swarm Intelligence and Bio-inspired Computation: Theory and Applications, pp. 169–191, Elsevier Science & Technology Books, 2013.

23.   O. Okobiah, S. P. Mohanty, and E. Kougianos, "Geostatistics inspired fast layout optimization of nanoscale CMOS phase locked loop," in Proceedings of the 14th International Symposium on Quality Electronic Design (ISQED '13), pp. 546–551, Santa Clara, Calif, USA, March 2013.

24.   W.-Y. Chang, "The state of charge estimating methods for battery: a review," ISRN Applied Mathematics, vol. 2013, Article ID 953792, 7 pages, 2013.

25.   T. Ganesan, I. Elamvazuthi, K. Z. Ku Shaari, and P. Vasant, "Swarm intelligence and gravitational search algorithm for multi-objective optimization of synthesis gas production," Applied Energy, vol. 103, pp. 368–374, 2013.

26.   O. T. Altinoz and A. E. Yilmaz, "Calculation of optimized parameters of rectangular patch antenna using gravitational search algorithm," in Proceedings of the International Symposium on INnovations in Intelligent SysTems and Applications (INISTA '11), pp. 349–353, IEEE, June 2011.

27.   Ö. T. Altinöz, A. E. Yilmaz, and G. W. Weber, "Orthogonal array based performance improvement in the gravitational search algorithm," Turkish Journal of Electrical Engineering and Computer Sciences, vol. 21, no. 1, pp. 174–185, 2013.

28.   N. M. Sabri, M. Puteh, and M. R. Mahmood, "A review of gravitational search algorithm," International Journal of Advances in Soft Computing and Its Applications, vol. 5, no. 3, pp. 1–39, 2013.

29.   R. E. Haskell, G. Castelino, and B. Mirshab, "Efficient algorithm for locating the global maximum of an arbitrary univariate function," Journal of Forth Application and Research, vol. 5, no. 3, pp. 357–364, 1989.

30.   M. Udgir, H. M. Dubey, and M. Pandit, "Gravitational search algorithm: a novel optimization approach for economic load dispatch," in Proceedings of the Annual International Conference on Emerging Research Areas and International Conference on

Microelectronics, Communications and Renewable Energy (AICERA/ICMiCR ‹13), pp. 1–6, 2013.

31. Y. Zheng, S. Chen, Y. Lin, and W. Wang, "Bio-inspired optimization of sustainable energy systems: a review," Mathematical Problems in Engineering, vol. 2013, Article ID 354523, 12 pages, 2013.

32. R. S. Barr, B. L. Golden, J. P. Kelly, M. G. C. Resende, and W. R. Stewart Jr., "Designing and reporting on computational experiments with heuristic methods," Journal of Heuristics, vol. 1, no. 1, pp. 9–32, 1995.

33. E. Cuevas, D. Zaldívar, and M. Pérez-Cisneros, "A swarm optimization algorithm for multimodal functions and its application in multicircle detection," Mathematical Problems in Engineering, vol. 2013, Article ID 948303, 22 pages, 2013.

34. A. Sadrnia, N. Ismail, N. Zulkifli, M. Ariffin, H. Nezamabadi-pour, and H. Mirabi, "A multiobjective optimization model in automotive supply chain networks," Mathematical Problems in Engineering, vol. 2013, Article ID 823876, 10 pages, 2013.

35. S. Mirjalili, S. Z. Mohd Hashim, and H. Moradian Sardroudi, "Training feedforward neural networks using hybrid particle swarm optimization and gravitational search algorithm," Applied Mathematics and Computation, vol. 218, no. 22, pp. 11125–11137, 2012.

Chapter 4

Bio-inspired Optimization of Sustainable Energy Systems: A Review

Yu-Jun Zheng[1], Sheng-Yong Chen[1], Yao Lin[2], and Wan-Liang Wang[1]

[1]College of Computer Science & Technology, Zhejiang University of Technology, Hangzhou 310023, China

[2]College of Life Sciences, Fujian Normal University, Fuzhou, Fujian 350108, China

ABSTRACT

Sustainable energy development always involves complex optimization problems of design, planning, and control, which are often computationally difficult for conventional optimization methods. Fortunately, the continuous advances in artificial intelligence have resulted in an increasing number of heuristic optimization methods for effectively handling those complicated problems. Particularly, algorithms that are inspired by the principles of natural biological evolution and/or collective behavior of social colonies have shown a promising performance and are becoming more and more popular nowadays. In this paper we summarize the recent advances in bio-inspired optimization methods, including artificial neural networks, evolutionary algorithms, swarm intelligence, and their hybridizations, which are applied to the field of sustainable energy development. Literature reviewed in this paper shows the current state of the art and discusses the potential future research trends.

INTRODUCTION

The demand for energy supply is increasing rapidly in recent years and will probably continue to grow in the future. The realization that fossil fuel resources are becoming scarce and that climate change is related to carbon emissions has stimulated interest in sustainable energy development [1]. In general, sustainable energy development strategies involve three major technological changes: energy savings on the demand side, efficiency improvements in the energy production, and replacement of fossil fuels by various sources of renewable energy [2]. In particular, due to its multifold advantages including inexhaustibility, safety, decrease in external energy dependence, decrease in impact of electricity production and transformation, increase in the level of services for the rural population, and so forth [3], renewable energy is now considered an important resource around the world and regarded as a key component in obtaining a sustainable development of our society.

The implementation of sustainable energy development strategies involves a wide range of design, planning, and control optimization problems. Various conventional optimization methods, such as linear programming [4–6], integer programming [7, 8], mixed integer linear programming [9–12], nonlinear programming [13–16], dynamic programming (DP) [17–20], constrained programming [21, 22], and so forth, have been applied for solving these problems. Nevertheless, current optimization problems in sustainable energy systems become more and more complex, especially when they include the integration of renewable sources in coherent energy systems. This is because most of such problems are nonlinear, nonconvex, with multiple local optima, and included in the category of NP-hard problems [23]. In consequence, those conventional methods might need exponential computation time in the worst case to obtain the optimum, which leads to computation time that is too high for practical purposes [24]. In recent years, modern heuristic optimization techniques, which are stochastic search methods inspired by the concepts and principles of artificial intelligence, have gained popularity in the optimization of sustainable energy systems.

In this paper, we give an overview of the latest research advances in bio-inspired solution methods for sustainable energy development. We particularly focus on the bio-inspired optimization algorithms that have been applied to the design, planning, and control problems in the field of renewable and sustainable energy systems. We roughly group those methods into three categories, which are artificial neural networks (ANNs), evolutionary algorithms (EAs), and swarm intelligence. Besides, we also describe the recent work about the hybridization of individual methods. These heuristic methods usually do not require deep mathematical knowledge and have been demonstrated to be quite useful and efficient in optimization search for complex optimization problems in science and engineering. We believe that this paper can help researchers to gain knowledge about the major developments emerged throughout the years and find valuable approaches that can be applied in the practice of implementing sustainable energy systems.

The rest of the paper is synthesized as follows: Section 2 reviews the application of ANNs in sustainable energy development, Section 3 summarizes the work about EAs applied to different types of optimization problems in energy, Section 4 presents the recent advances in swarm-based methods used in the field, Section 5 introduces the hybrid techniques combining two or more of above methods, and Section 6 concludes with discussion.

ARTIFICIAL NEURAL NETWORKS

An ANN is a collection of neuron-like processing units with weighted connections between the units, which is inspired by our present understanding of biological nervous systems. Roughly speaking, ANNs use processing elements connected by links of variable weights to form a black box representation of systems [25]. ANNs can be trained by adjusting the weights so as to be able to predict or classify new patterns, and they provide some of the human characteristics of problem solving that are difficult to simulate using other computational technologies.

Advantages of ANN include their high tolerance of noisy data, their ability to process patterns on which they have not been trained, as well as that they can be used without much preliminary knowledge about the problem domain. However, ANNs typically involve long training times and have been criticized for their poor interpretability [26].

ANNs are popular for prediction and forecasting nonlinear physical series (such as wind [27] and water lever [28]) which are beyond the capability of linear predictors such as autoregressive (AR), moving average (MA), and autoregressive moving average (ARMA), [29–31]. Since 1990s, various studies have been reported on the applications of ANN in predicting electric loads and energy demands. An early work of Kawashima [32] developed an ANN backpropagation model with three-phase annealing for the first building energy prediction competition held by the American Society of Heating, Refrigerating- and Air-Conditioning Engineers in 1993. Islam et al. [33] proposed an ANN-based weather-load and weather-energy models, where a set of weather and other variables are identified for both models together with their correlations and contribution to the forecasted variables. They applied the models to historical energy, load, and weather data available for the Muscat power system from 1986 to 1990, and the forecast results for 1991-1992 show that monthly electric energy and load can be predicted within a maximum error of 6% and 10%. Al-Shehri [34] used an ANN model for forecasting the residential electrical energy in the Eastern Province of Saudi Arabia, the forecasting result of which is shown to be closer to the real data than that predicted by the polynomial fit model. Azadeh et al. [35] developed a simulated-based ANN and applied it in forecasting monthly electrical energy consumption in Iran from March 1994 to February 2005 (131 months), and the result shows that the ANN model always provides the best solutions and estimation in comparison with other models such as time series.

ANNs have also been applied in midterm and long-term energy forecasting in different industrial sectors, areas, and countries and demonstrated their superiorities in comparison with conventional prediction models [36–41]. In [42] Ermis et al. presented an ANN model which is trained based on world energy consumption data from 1965 to 2004 and applied for forecasting world green energy consumption to the year 2050. It is estimated that world green energy and natural gas consumption will continue increasing after 2050, while world oil and coal consumption are expected to remain relatively stable after 2025 and 2045, respectively.

In recent years, ANN-based models have also been widely used in design and implementation of different kinds of renewable energy systems. For example, in the design of solar energy systems the estimation and calculation of radiation data are very important. Bosch et al. [43] presented an ANN approach for calculating solar radiation levels over complex mountain terrains using data from only one radiometric station. Cao and Lin [44] proposed a diagonal recurrent wavelet neural network which uses historical information of cloud cover to sample data sets for network training and applied their approach in hourly irradiance forecasting in Shanghai, China. Zervas et al. [45] developed an ANN-based prediction model of global solar irradiance distribution on horizontal surfaces, which has been applied to the meteorological database of NTUA, Zografou Campus, Athens.

In the same manner, the prediction of water level is fundamental for ocean energy generation. Huang et al. [46] developed an ANN for water level predictions, which has been applied to coastal inlets taking into account long-term water level observations. Kazeminezhad et al. [47] studied an ANN-based fuzzy inference system for predicting wave parameters, with an application to the data set comprising of fetch-limited wave data and over water wind data gathered from deep-water location in Lake Ontario.

The performance of photovoltaic system heavily depends on the meteorological conditions, and sizing represents an important part of photovoltaic systems design, that is, the optimal selection of the number of solar, cell panels, the size of the storage battery, and the size of wind generator to be used for certain hybrid applications [48]. ANNs have the capability to model complex, nonlinear processes without having to assume the form of the relationship between input and output variables, and thus ANN-based models, including adaptive ANN [49, 50], recurrent ANN [51], radial basis function network (RBFN) [52], have been successfully applied for sizing of photovoltaic systems.

EVOLUTIONARY ALGORITHMS

Evolutionary algorithms (EAs) are stochastic search methods inspired by the principles of natural biological evolution for computationally difficult problems. They are very suitable for complex engineering optimization problems which may be multimodal, nondifferentiable, or discontinuous and thus cannot be solved by conventional gradient-based methods. In general, An EA simultaneously evolves a population of possible solutions and also returns a population of solutions. Typical EAs include genetic algorithms (GAs) [53], evolutionary programming (EP) [54], evolution strategies (ES) [55], differential evolution (DE) [56], and biogeography-based optimization (BBO) [57]. The advantages of EAs include their relative simplicity of implementation, inherent parallel architecture, and scalability to high-dimensional solution spaces.

Moreover, in real-world applications there are a large number of multiobjective optimization problems, that is, problems requiring the simultaneous optimization of several objectives which are often conflicted. For most of such problems, there is no single optimal solution and thus a solution method should search for a set of nondominated (Pareto optimal) solutions, that is, all the solutions such that there exists no other individual better in all the objectives. EAs are capable of finding several members of the Pareto optimal set in a single run of the algorithm, instead of having to perform a series of separate runs as in the case of the traditional mathematical programming techniques [58] and thus are very suitable for tackling with complex multiobjective optimization problems.

Genetic Algorithms

Genetic algorithms (GAs) are of the famous evolutionary algorithms which simulate the Darwinian principle of natural selection and the survival of the fittest in optimization [53]. A GA typically works with a fixed-size population of solutions and uses three genetic operations, namely selection, crossover, and mutation, to modify the solutions chosen from the current generation and select the most appropriate offspring to pass on to the next generations.

A number of researches have been reported on the application of GA in the optimal design and operation of sustainable energy systems.

For wind energy systems, Li et al. [59] used a multilevel GA to solve the optimal design problem of integrating the number of actuators, the configuration of the actuators, and the active control algorithms in buildings excited by strong wind force. Li et al. [60] employed a GA to optimize the ranges of gearbox ratios and power ratings of multihybrid permanent-magnet wind generator systems. Grady et al. [61] used a GA to determine the optimal placement of wind turbines for maximum production capacity while limiting the number of turbines installed and the acreage of land occupied by each wind farm. Emami and Noghreh [62] proposed a GA with a new coding and a new objective function with adjustable coefficients for the similar problem, and their algorithm shows better performance on the optimal control of the cost, power, and efficiency of the wind farm. For solar energy systems, Varun and Siddhartha [63] proposed a GA to optimize system parameters in order to maximize the thermal performance of flat plate solar air heaters. Zagrouba et al. [64] adapted a GA to identify the electrical parameters of photovoltaic solar cells and modules to determine the maximum power point from the illuminated current-voltage characteristic. GAs have also been used in geothermal systems [65] and hybrid photovoltaic systems [66–70].

Evolutionary Programming and Evolution Strategies

Evolutionary Programming (EP) was devised in order to evolve finite state machines for the prediction of events on the basis of former observations and has been demonstrated useful for searching the optimum of nonlinear functions [71]. Cau and Kaye [72] proposed a constructive EP approach to minimize the cost of operating a power system with multiple distributed energy storage resources. Their approach combines DP and EP by evolving piecewise linear convex cost-to-go functions and thus decomposes the multistage scheduling problem into smaller one-stage sub-problems which are easy to cope with. Fong et al. [73] developed a simulation-EP coupling method to solve the discrete, nonlinear, and highly constrained optimization problems related to energy management of heating, ventilating, and air-conditioning (HVAC) systems. The application of the method to a local HVAC installation project achieved a saving potential of about 7% as compared to the existing operational settings, without any extra cost. In [74] MacGill presented a dual EP approach integrating with software agents for power system resources to coevolve optimal operational behaviors over repeated power system simulations. The proposed tool was successfully applied to a real-world problem exploring the potential operational synergies between significant PV penetrations and distributed energy storage options including controllable loads.

Evolution strategies (ES) are a class of general optimization methods which evolve a population of solutions by means of variation and selection. Original ES uses a mutation operator that produces a single descendent from a given ancestor, denominated ES-(1+1), and was progressively generalized to ES-($\mu+\lambda$), that is, several ancestors ($\mu > 1$) and descendents ($\lambda > 1$) in each generation [75]. In [76] Chang used an ES approach to solve optimal chiller loading problem, which takes the chilled water supply temperature as the variable to be determined for the decoupled air-conditioning system. The result shows, the approach outperforms both the Lagrangian method and the GA method. Considering the optimal selection and sizing of distributed energy resources which can be formulated as a nonlinear mixed-integer minimization problem, Logenthiran et al. [77] used ES for the minimization of capital and annual operational cost of DER under a variety of system and unit constraints. Their method was applied to design integrated microgrids for an intelligent energy distribution system project.

Differential Evolution

Differential evolution (DE) approach combines simple arithmetic operators with the classical operators of crossover, mutation, and selection to evolve a randomly generated starting population to a final solution. It is similar to a ($\mu + \lambda$) ES, but in DE the mutation is not done via some separately defined probability density function [78]. Chakraborty et al. [79] presented a fuzzy DE method for solving thermal unit commitment problem integrated with solar energy system, where the solar radiation, forecasted load demand and associated constraints are formulated as fuzzy sets considering the error. Slimani and Bouktir [80] developed a DE method to solve the optimal power flow problem, whose objective function is the minimization of the cost of the thermal and the wind generators with different sizes. The method decomposes the optimization constraints of the power system into active constraints manipulated directly by DE, and passive constraints maintained in their soft limits using a conventional constraint load flow.

dos Santos Coelho et al. [81] developed a cultural DE algorithm for optimizing the economic dispatch of electrical energy using thermal generators and validated their approach on a test system consisting of 13 thermal generators whose nonsmooth fuel cost function takes into account the valve-point loading effects. Suzuki et al. [82] studied a large-scale mixed-integer nonlinear problem for generating optimal operational planning for energy plants and developed an constrained DE algorithm to effectively solve the problem without much parameters tuning effort. Hejazi et al. [83] developed a DE algorithm for optimal allocation of energy and spinning reserve, taking all security and power systems constraints in steady state and system credible contingencies into consideration. Lee et al. [84] conducted a comparative study of DE, GA, PSO, and LP methods for solving the optimal chiller loading problem for reducing energy consumption, and the result shows that the DE algorithm achieves the best result. Peng et al. [85] considered a problem in the design of the Earth-Moon low-energy transfer to find the patch point of the unstable manifold of the Lyapunov orbit around Sun-Earth L2 and the stable manifold of the Lyapunov orbit around Earth-Moon L2. They designed an improved differential evolution algorithm which incorporates the uniform design technology and the self-adaptive parameter control method into standard differential evolution to accelerate its convergence speed and improve the stability, and thus effectively solve the problem.

Multiobjective Evolutionary Algorithms

Multiobjective evolutionary algorithms (MOEAs) have received much interest in recent years. A number of metaheuristic algorithms, such as the nondominated sorting genetic algorithm NSGA [86] and the NSGA-II [87], the strength Pareto evolutionary algorithm (SPEA) [88] and the SPEA2 [89], the Pareto archived evolution strategy (PAES) [90], the Pareto differential evolution algorithm (PDE) [91], the nondominated sorting differential evolution (NSDE) [92], and so forth have gained great success in solving multiobjective optimization problems [93].

Benini and Toffolo [94] presented an MOEA for the design of stall-regulated horizontal-axis wind turbines, the aim of which is to achieve the best trade-off performance between the total energy production per square meter of wind park and cost. Their method can optimize the geometrical parameters of the rotor configuration of wind turbines, achieving the best trade-off performance between the two objectives. Zhao et al. [95] employed a GA whose input parameters are the main components of a wind farm and key technical specifications and whose output is an optimal electrical system design of the wind farm which is optimized in terms of both production cost and system reliability. Kusiak et al. [96] proposed an MOEA for evaluating wind turbine performance, where the objectives include the maximization of the wind power output and the minimization of the vibration of the drive train and of the tower. In [97] Kusiak and Song used the MOEA for optimizing wind turbine placement based on wind distribution, including the selection the best turbine combination from a given list of available turbines.

Bernal-Agustín et al. [98] applied the SPEA to the design of a photovoltaic-wind-diesel system, where the objectives include the minimization of both the total cost throughout the useful life of the installation and the pollutant emissions. They later applied the algorithm to an extension of the problem, which adds an objective of the unmet load in the system [99]. Ould Bilal [100] proposed a multiobjective GA for minimizing the annualized cost system and the loss of power supply probability of a hybrid solar-wind-battery system. Montoya et al. [101] combined PAES with simulated annealing (SA) and tabu search (TS) to minimize voltage deviations and power losses in power networks. Thiaux et al. [102] applied NSGA-II to optimize stand-alone photovoltaic systems by reducing the gross energy requirement and minimizing the storage capacity. In [103] Rao and Peng considered a multiobjective optimal model of dispatch of energy-saving and emission reduction generation in the power system and developed a multiobjective DE algorithm with niche strategy for improving the crowing mechanism in the process of Pareto nondominated sorting operation. The experiment shows that their method can achieve better result than NSGA-II and NSDE.

SWARM INTELLIGENCE

The expression "swarm intelligence" was originally used in the context of cellular robotic systems to describe the self-organization of simple mechanical agents through nearest-neighbor interaction [104]. Bonabeau et al. [105] extended the definition to include "any attempt to design algorithms or distributed problem-solving devices inspired by the collective behavior of social insect colonies and other animal societies." Since the 1990s, a number of swarm-based algorithms, including particle swarm optimization (PSO) [106], ant colony optimization (ACO) [107], artificial bee algorithms [108, 109], artificial immune systems (AIS) [110] have been proposed for difficult optimization problems especially with large continuous or combinatorial search spaces.

Particle Swarm Optimization

PSO is another population-based global optimization technique that enables a number of individual solutions, called particles, to move through a hyperdimensional search space to search for the optimum. Each particle has a position vector and a velocity vector, which are adjusted at iterations by learning from a local best found by the particle itself and a current global best found by the whole swarm. Empirical studies have shown that PSO has a high efficiency in convergence to desirable optima and performs better than GA and other EAs on many problems [111].

AlRashidi and EL-Naggar [112] employed a PSO algorithm for estimating annual peak load forecasting in an electrical power system, the aim of which is to minimize the error associated with the estimated model parameters. Their approach was validated on actual recorded data from Kuwaiti and Egyptian networks. Niknam and Firouzi [113] developed a PSO algorithm combined with simplex search, for estimating load and renewable energy source output on the power systems, and their comparative experiment show that the PSO performs better than several EAs and other swarm-based algorithms.

Amjady and Soleymanpour [114] developed a modified adaptive PSO for daily hydrothermal generation scheduling, which is a complicated nonlinear, nonconvex, and nonsmooth optimization problem with discontinuous solution space. As some other adaptive PSOs [115, 116], their method dynamically changes the inertia weight and acceleration coefficients of the algorithm to increase activities of particles to explore broad space. Lee [117] applied PSO to solve short-term hydroelectric generation scheduling of a power system with wind turbine generators. Kongnam and Nuchprayoon [118] used PSO for the control problem of a wind turbine, which involves the determination of rotor speed and tip-speed ratio to maximize power and energy capture from the wind. Khanmohammadi et al. [119] developed a method based on PSO and Nelder-Mead algorithms for determining the optimal unit commitment (startups and shutdowns scheduling) of hydropower plants. López et al. [120] presented a binary PSO-based method to accomplish optimal location of biomass-fuelled systems for distributed power generation with forest residues as biomass source, and the results outperformed those obtained by a GA when maximizing a profitability index taking into account technical constraints. In [121] the authors also applied a PSO algorithm for the optimal location and supply area for biomass-based power plants. There are also a number of researches reported on the application of PSO in the design and control of hybrid photovoltaic systems [122–126].

Economic dispatch problems, the main aim of which is to schedule the committed generating units output so as to meet the required load demand at minimum cost satisfying all unit and system operational constraints, typically have nonlinear, nonconvex type objective function with intense equality and inequality constraints. Mahor et al. [127] presented a yearly (2003–2008) review of work of application of PSO to solve the various economic dispatch problems. The algorithms include linearly varying inertia weight PSO [128, 129], PSO with constriction factor and inertia weight [130, 131], PSO with linearly varying inertia weight with constriction factor [132], chaotic PSO [133–135], and multiobjective PSO [136–139]. It was suggested that PSO algorithms (in particular those with time varying control parameters) can give an improved results within less computational time in comparison to conventional methods, but still further improvements in PSO algorithms are required, especially for real-time scheduling problems.

Ant Colony Optimization

Ant colony optimization (ACO) algorithms mimic the behavior of real ants living in colonies that communicate with each other using pheromones in order to accomplish complex tasks such as establishing a shortest path from the nest to food sources [107]. Li et al. [140] applied an ACO algorithm to the optimal design of solar energy dynamic power system in space station, with the aim to minimize the launching mass of the system subject to a set of constraints on parameters including pressure, temperature, compression coefficient, numbers and diameter of heat exchangers, height of recycling refrigerant, and so forth. Considering the optimal sizing of the design of standalone hybrid wind/photovoltaic power systems, Xu et al. [141] used ACO to minimize the total capital cost, subject to the constraint of the loss of power supply probability calculated by simulation. Foong et al. [142] considered a power plant maintenance scheduling optimization formulation incorporating the options of shortening the maintenance duration and/or deferring maintenance tasks in the search for practical maintenance schedules and developed an improve ACO algorithm for solving the problem. Warner and Vogel [143] considered planning of an energy supply network by simultaneously choosing the plants and the optimal network and implemented an ACO algorithm for the problem. See et al. [144] used ACO for determining optimal parameter values to the control model of energy extraction and thus improving the performance of wave energy converters as well as their long-term economic value.

Toksari [145] proposed an ACO electricity energy estimation model for forecasting electricity energy generation and demand, taking population, gross domestic product (GDP), import and export into consideration. He found that the model with quadratic equations can provide better fit solution due to the fluctuations of the economic indicators. The proposed model was applied to indicate Turkey's net electricity energy generation and demand until 2025. Baskan et al. [146] used ACO for estimating the transport energy demand of Turkey using gross domestic product, population, and vehicle-km. It is also expected that the work will be helpful in developing highly applicable and productive planning for transport energy policies.

Artificial Bee Algorithms

Artificial bee algorithms simulate the intelligent foraging behavior of a honeybee swarm. Two most popular algorithms are the artificial bee colony (ABC) algorithm and the honey bee mating optimization (HBMO) algorithm [147]. Niknam et al. [148] presented a multiobjective HBMO algorithm for the siting and sizing of renewable electricity generators, in order to optimize the placement of renewable electricity generators by considering objective functions including losses, costs of electrical generation, and voltage deviation. In [149] Niknam et al. also proposed an improve a HBMO algorithm for economic dispatch in power systems, with the aim to get maximum usable power using minimum resources. Abu-Mouti and El-Hawary [150] considered a dynamic economic dispatch problem, whose aim is to determine the optimal power outputs of online generating units in order to meet the load demand subject to satisfying various operational constraints over finite dispatch periods, and they applied an ABC algorithm to solve the problem.

Vera et al. [151] proposed a binary honey bee foraging (HBF) swarm approach for searching the optimal location, biomass supply area, and power plant size that offer the best profitability for investor. Experimental results show that the HBF approach method outperforms PSO and GA. Hong [152] presented an electric load forecasting model based on a chaotic ABC algorithm combined with the seasonal recurrent support vector regression model, and the experiments indicated that the model can provide a promising forecasting performance for electric load.

Artificial Immune System (AIS)

Inspired by the theoretical immunology, observed immune functions, principles, and models, AIS stimulates the adaptive immune system of a living creature to unravel the various complexities in real-world engineering optimization problems. Abdul Rahman et al. [153] developed an AIS algorithm for the economic dispatch problem, which uses the total generation cost as the objective function. Through genetic evolution, the antibodies with high affinity measure are produced and become the solution, and the algorithm converges within an acceptable execution time and highly optimal solution for economic dispatch with minimum generation cost. Coelho and Mariani [154] coped with the problem by using a chaotic artificial immune network approach, which has been demonstrated by the experiments to be an effective alternative to schedule the committed generating unit outputs to meet the required load demand at minimum operating cost while satisfying system constraints. Recently, Arsalani and Seddighizadeh [155] used an AIS algorithm to minimize the deviation of bus voltage from its nominal value as well as the loss of energy in a power system. The main advantage of the algorithm is that it prevents many times repetition of similar solutions, and the result shows that the algorithm can achieve a solution that meets a level of preferences better than that required although the threshold is determined by means of fuzzy logic to reflect the imprecise nature of optimization objectives.

HYBRID METHODS

By exploiting the advantages and disadvantages of two or more solution methods, we have a chance to obtain a powerful approach that is much more competitive than any individual method. Research and development on hybrid bio-inspired methods in sustainable energy systems have grown dramatically since the late 1990s.

Mellit and Kalogirou [156] studied the combination of GA and ANN for optimal sizing of stand-alone photovoltaic systems. Firstly the GA was used to optimize the sizing parameters of sites, and then the ANN was used to predict the optimal parameters in remote areas. Mellit later developed a hybrid model combining adaptive-network-based fuzzy inference system (ANFIS) and GA and demonstrated that the model with ANFIS presents more accurate results [157]. Chang and Ko [158] designed a hybrid heuristic method which combines PSO with nonlinear time-varying evolution and ANN in order to determine the tilt angle of photovoltaic modules with the aim of maximizing the electrical energy output of the modules.

Li et al. [159] proposed a method combining AIS and PSO, for optimal load distribution among cascade hydropower stations. Their hybrid method involves the immune information processing mechanism into PSO and thus improves the ability to find the globally excellent result and the convergence speed with its special concentration selection mechanism and immune vaccination. Yang et al. [160] combined GA and ABC into a bee evolutionary genetic algorithm (BEGA), which has characteristics of higher precision and faster convergence rate and has been effectively applied to a problem of minimizing the energy consumption of central air-conditioning system without lowering the degree of comfort. The test on a common load distribution case shows that the hybrid method can achieve an energy-saving rate at 25.1%.

Kıran et al. [161] proposed a hybrid method of PSO and ACO for estimating energy demand, PSO for solving continuous optimization part and ACO for discrete part. The experiments demonstrated that the hybrid method outperforms both the individual PSO and ACO. In [162] Ghanbari et al. combined GA and ACO to model and simulate fluctuations of energy demand under the influence of related factors. Firstly the GA is used for generating data base of the expert system, and then the ACO is used for learning linguistic fuzzy rules such that degree of cooperation between data base and rule base increases. Results showed that the method can provide more accurate-stable results than ANFIS- and ANN-based approaches.

DISCUSSION AND CONCLUSIONS

We have summarized the recent research advances in bio-inspired solutions applied to the design, control, and implementation of sustainable energy systems. Typical illustrations are addressed for ANNs, EAs, swarm-based algorithms, and their hybridizations. Representative works are summarized to help readers have a general overview of the state-of-the-art and easily refer suitable methods in practical solutions.

The first finding of this paper is that the number of research papers on bioinspired optimization algorithms on sustainable energy problems has increased dramatically since 1990s. A large percent of early work was GA related. However, in recent years, DE has become more popular in the category of EAs, and swarm-based methods have gained more and more attentions of the researchers and practitioners. In the last three years, we found that PSO algorithms have become one of the most widely used methods in the field of renewable and sustainable energy development.

In general, none of the individual methods could perform better than all the other methods on all kinds of problems, suggesting that customized methods need to be carefully chosen or designed according to the respective problem. But researchers and practitioners can learn from the experiences of early researchers. For example, on most problems of unit sizing of stand-alone hybrid energy systems, PSO typically outperforms GAs [163], mainly because PSO algorithms are more suitable for high-dimensional optimization problems, and improved versions of PSO are less sensitive to multiple local optima than GAs.

With the increasing importance and complexity of energy systems, we are facing the challenges to promote the performance, reliability, and scalability of solution methods [164, 165]. In consequence, it can be anticipated that future research will continuously put great emphasis on the hybridization of bio-inspired methods. In addition, more and more real-world problems in sustainable energy consider more than one objective. It can be expected that multiobjective bio-inspired optimization algorithms and parallel processing will be promising research areas in this field [166]. Moreover, current studies on multiobjective algorithms combing more than one metaheuristics are still rare, and we think that this can be a valuable direction for the researchers.

Today's new computational paradigms, such as quantum computing [167], DNA computing [168], and fractal computing [169–172], provide valuable inspiration for creating new heuristics for extremely difficult problems. Thus, the extensions of current bio-inspired methods based on these new paradigms are expected to achieve dramatic improvement on computational performance. For example, quantum-inspired EAs are regarded as one of the three main research areas related to the complex interaction between quantum computing and EAs [173]. In the aspect of quantum computing, if applying ANNs, it is worth considering time series models in that aspect as that discussed by Bakhoum and Toma [174, 175]. We believe that the fruits of these researches are continuously becoming new technological solutions to new open problems, and the full potential is far from being reached.

ACKNOWLEDGMENTS

This work was supported by the National Natural Science Foundation of China (61105073, 61173096, and 61103140), Doctoral Fund of Ministry of Education of China (20113317110001), and Zhejiang Provincial Natural Science Foundation (R1110679).

REFERENCES

1. E. Vine, "Breaking down the silos: the integration of energy efficiency, renewable energy, demand response and climate change," Energy Efficiency, vol. 1, no. 1, pp. 49–63, 2008. · ·

2. H. Lund, "Renewable energy strategies for sustainable development," Energy, vol. 32, no. 6, pp. 912–919, 2007. · ·

3. A. Hepbasli, "A key review on exergetic analysis and assessment of renewable energy resources for a sustainable future," Renewable and Sustainable Energy Reviews, vol. 12, no. 3, pp. 593–661, 2008. · ·

4. R. Chedid and Y. Saliba, "Optimization and control of autonomous renewable energy systems,"International Journal of Energy Research, vol. 20, no. 7, pp. 609–624, 1996.

5. S. Iniyan and K. Sumathy, "The application of a Delphi technique in the linear programming optimization of future renewable energy options for India," Biomass and Bioenergy, vol. 24, no. 1, pp. 39–50, 2003. · ·

6. G. Privitera, A. R. Day, G. Dhesi, and D. Long, "Optimising the installation costs of renewable energy technologies in buildings: a linear programming approach," Energy and Buildings, vol. 43, no. 4, pp. 838–843, 2011. · ·

7. M. Gong, "Optimization of industrial energy systems by incorporating feedback loops into the MIND method," Energy, vol. 28, no. 15, pp. 1655–1669, 2003. · ·

8. P. Liu, D. I. Gerogiorgis, and E. N. Pistikopoulos, "Modeling and optimization of polygeneration energy systems," Catalysis Today, vol. 127, no. 1–4, pp. 347–359, 2007. · ·

9. T. Ikegami, Y. Iwafune, and K. Ogimoto, "Development of the optimum operation scheduling model of domestic electric appliances for the supply-demand adjustment in a power system," IEEJ Transactions on Power and Energy, vol. 130, no. 10, pp. 877–887, 2010. · ·

10. B. Wille-Haussmann, T. Erge, and C. Wittwer, "Decentralised optimisation of cogeneration in virtual power plants," Solar Energy, vol. 84, no. 4, pp. 604–611, 2010. · ·

11. H. Morais, P. Kádár, P. Faria, Z. A. Vale, and H. M. Khodr, "Optimal scheduling of a renewable micro-grid in an isolated load area using mixed-integer linear programming," Renewable Energy, vol. 35, no. 1, pp. 151–156, 2010. · ·

12. S. Ruangpattana, D. Klabjan, J. Arinez, and S. Biller, "Optimization of on-site renewable energy generation for industrial sites," in Proceedings of IEEE/PES Power Systems Conference and Exposition (PSCE ‹11), March 2011. · ·

13. C. A. Babu and S. Ashok, "Optimal utilization of renewable energy-based IPPs for industrial load management," Renewable Energy, vol. 34, no. 11, pp. 2455–2460, 2009. · ·

14. Z. Kravanja, "Mathematical programming approach to sustainable system synthesis," Chemical Engineering Transactions, vol. 21, pp. 481–486, 2010.

15. A. Borghetti, M. Bosetti, S. Grillo et al., "Short-term scheduling and control of active distribution systems with high penetration of renewable resources," IEEE Systems Journal, vol. 4, no. 3, pp. 313–322, 2010. · ·

16. A. Vergnano, C. Thorstensson, B. Lennartson, et al., "Modeling and optimization of energy consumption in cooperative multi-robot systems," IEEE Transactions on Automation Science and Engineering, vol. 9, no. 2, pp. 423–428, 2012. ·

17. E. D. Castronuovo and J. A. P. Lopes, "Optimal operation and hydro storage sizing of a wind-hydro power plant," International Journal of Electrical Power and Energy System, vol. 26, no. 10, pp. 771–778, 2004. · ·

18. N. Löhndorf and S. Minner, "Optimal day-ahead trading and storage of renewable energies—an approximate dynamic programming approach," Energy Systems, vol. 1, no. 1, pp. 61–77, 2010. · ·

19. V. Marano, G. Rizzo, and F. A. Tiano, "Application of dynamic programming to the optimal management of a hybrid power plant with wind turbines, photovoltaic panels and compressed air energy storage," Applied Energy, vol. 97, pp. 849–859, 2012.

20. A. Sinha and P. Chaporkar, "Optimal power allocation for a renewable energy source," in Proceedings of National Conference on Communications, 2012.

21. Y. P. Cai, G. H. Huang, Z. F. Yang, Q. G. Lin, and Q. Tan, "Community-scale renewable energy systems planning under uncertainty—an interval chance-constrained programming approach," Renewable and Sustainable Energy Reviews, vol. 13, no. 4, pp. 721–735, 2009. · ·

22. S. Tomasin and T. Erseghe, "Constrained optimization of local sources generation in smart grids by SDP approximation," in Proceedings of IEEE International Symposium on Power Line Communications and Its Applications (ISPLC ‹11), pp. 187–192, April 2011. · ·

23. M. R. Garey and D. S. Johnson, Computers and Intractability: A Guide to the Theory of NP-Completeness, A Series of Books in the Mathematical Sciences, W. H. Freeman, San Francisco, Calif, USA, 1979. ·

24. R. Baños, F. Manzano-Agugliaro, F. G. Montoya, C. Gil, A. Alcayde, and J. Gómez, "Optimization methods applied to renewable and sustainable energy: a review," Renewable and Sustainable Energy Reviews, vol. 15, no. 4, pp. 1753–1766, 2011.

25. K. Chau, "A review on the integration of artificial intelligence into coastal modeling," Journal of Environmental Management, vol. 80, no. 1, pp. 47–57, 2006.

26. J. Han, M. Kamber, and J. Pei, Data Mining: Concepts and Techniques, Morgan Kaufmann, Waltham, Mass, USA, 3rd edition, 2012.

27. M. Li, S. C. Lim, and W. Zhao, "Cauchy-Matern model of sea surface wind speed at the Lake Worth, Florida," Mathematical Problems in Engineering, vol. 2012, Article ID 843676, 10 pages, 2012. ·

28. M. Li, Y. Q. Chen, J.-Y. Li, and W. Zhao, "Hölder scales of sea level," Mathematical Problems in Engineering, vol. 2012, Article ID 863707, 22 pages, 2012. ·

29. J. Makhoul, "Linear prediction: a tutorial review," Proceedings of the IEEE, vol. 63, no. 4, pp. 561–580, 1975.

30. B. S. Atal, "The history of linear prediction," IEEE Signal Processing Magazine, vol. 23, no. 2, pp. 154–161, 2006.

31. M. Li, W. Zhao, and B. Chen, "Heavy-tailed prediction error: a difficulty in predicting biomedical signals of 1/f noise type," Computational and Mathematical Methods in Medicine, vol. 2012, Article ID 291510, 5 pages, 2012. ·

32. M. Kawashima, "Artificial neural network backpropagation model with three-phase annealing developed for the building energy predictor shootout," ASHRAE Transactions, vol. 100, no. 2, pp. 1096–1103, 1994.

33. S. M. Islam, S. M. Al-Alawi, and K. A. Ellithy, "Forecasting monthly electric load and energy for a fast growing utility using an artificial neural network," Electric Power Systems Research, vol. 34, no. 1, pp. 1–9, 1995.

34. A. Al-Shehri, "Artificial neural network for forecasting residential electrical energy," International Journal of Energy Research, vol. 23, no. 8, pp. 649–659, 1999.

35. A. Azadeh, S. F. Ghaderi, and S. Sohrabkhani, "A simulated-based neural network algorithm for forecasting electrical energy consumption in Iran," Energy Policy, vol. 36, no. 7, pp. 2637–2644, 2008. · ·

36. G. J. Tsekouras, N. D. Hatziargyriou, and E. N. Dialynas, "An optimized adaptive neural network for annual midterm energy forecasting," IEEE Transactions on Power Systems, vol. 21, no. 1, pp. 385–391, 2006. · ·

37. A. Sözen, M. A. Akçayol, and E. Arcaklioğlu, "Forecasting net energy consumption using artificial neural network," Energy Sources, Part B, vol. 1, no. 2, pp. 147–155, 2006.

38. A. Azadeh, S. F. Ghaderi, and S. Sohrabkhani, "Annual electricity consumption forecasting by neural network in high energy consuming industrial sectors," Energy Conversion and Management, vol. 49, no. 8, pp. 2272–2278, 2008. · ·

39. A. H. Neto and F. A. S. Fiorelli, "Comparison between detailed model simulation and artificial neural network for forecasting building energy consumption," Energy and Buildings, vol. 40, no. 12, pp. 2169–2176, 2008. · ·

40. R. Yokoyama, T. Wakui, and R. Satake, "Prediction of energy demands using neural network with model identification by global optimization," Energy Conversion and Management, vol. 50, no. 2, pp. 319–327, 2009. · ·

41. Z. W. Geem and W. E. Roper, "Energy demand estimation of South Korea using artificial neural network," Energy Policy, vol. 37, no. 10, pp. 4049–4054, 2009.

42. K. Ermis, A. Midilli, I. Dincer, and M. A. Rosen, "Artificial neural network analysis of world green energy use," Energy Policy, vol. 35, no. 3, pp. 1731–1743, 2007.

43. J. L. Bosch, G. López, and F. J. Batlles, "Daily solar irradiation estimation over a mountainous area using artificial neural networks," Renewable Energy, vol. 33, no. 7, pp. 1622–1628, 2008. · ·

44. J. Cao and X. Lin, "Application of the diagonal recurrent wavelet neural network to solar irradiation forecast assisted with fuzzy technique," Engineering Applications of Artificial Intelligence, vol. 21, no. 8, pp. 1255–1263, 2008. · ·

45. P. L. Zervas, H. Sarimveis, J. A. Palyvos, and N. C. G. Markatos, "Prediction of daily global solar irradiance on horizontal surfaces based on neural-network techniques," Renewable Energy, vol. 33, no. 8, pp. 1796–1803, 2008. · ·

46. W. Huang, C. Murray, N. Kraus, and J. Rosati, "Development of a regional neural network for coastal water level predictions," Ocean Engineering, vol. 30, no. 17, pp. 2275–2295, 2003. · ·

47. M. H. Kazeminezhad, A. Etemad-Shahidi, and S. J. Mousavi, "Application of fuzzy inference system in the prediction of wave parameters," Ocean Engineering, vol. 32, no. 14-15, pp. 1709–1725, 2005. · ·

48. A. Mellit, S. A. Kalogirou, L. Hontoria, and S. Shaari, "Artificial intelligence techniques for sizing photovoltaic systems: a review," Renewable and Sustainable Energy Reviews, vol. 13, no. 2, pp. 406–419, 2009. · ·

49. A. Mellit, M. Benghanem, A. H. Arab, and A. Guessoum, "Modelling of sizing the photovoltaic system parameters using artificial neural network," in Proceedings of IEEE Conference on Control Applications, vol. 1, pp. 353–357, June 2003.

50. A. Mellit, M. Benghanem, A. H. Arab, and A. Guessoum, "An adaptive artificial neural network model for sizing stand-alone photovoltaic systems: application for isolated sites in Algeria," Renewable Energy, vol. 30, no. 10, pp. 1501–1524, 2005. · ·

51. L. Hontoria, J. Aguilera, and P. Zufiria, "A new approach for sizing stand alone photovoltaic systems based in neural networks," Solar Energy, vol. 78, no. 2, pp. 313–319, 2005. · ·

52. A. Mellit, M. Benghanem, A. Hadj Arab, and A. Guessoum, "Identification and modeling of the optimal sizing combination of stand-alone photovoltaic systems using the radial basis function networks," in Proceedings of the World Renewable Energy Congress VIII (WREC ‹04), Denver, Colo, USA, 2004.

53. J. H. Holland, Adaptation in Natural and Artificial Systems, University of Michigan Press, Ann Arbor, Mich, USA, 1975. ·

54. D. B. Fogel, "Introduction to simulated evolutionary optimization," IEEE Transactions on Neural Networks, vol. 5, no. 1, pp. 3–14, 1994. · ·

55. I. Rechenberg, Evolutions Strategies: Optimierung Technischer Systemenach Prinzipien der Biologischen Evolution, Frommberg-Holzboog, Stuttgart, Germany, 1973.

56. R. Storn and K. Price, "Differential evolution—a simple and efficient heuristic for global optimization over continuous spaces," Journal of Global Optimization, vol. 11, no. 4, pp. 341–359, 1997. · · · ·

57. D. Simon, "Biogeography-based optimization," IEEE Transactions on Evolutionary Computation, vol. 12, no. 6, pp. 702–713, 2008. · ·

58. K. Miettinen, Nonlinear Multiobjective Optimization, International Series in Operations Research & Management Science, 12, Kluwer Academic Publishers, Boston, Mass, USA, 1999.

59. Q. S. Li, D. K. Liu, J. Q. Fang, and C. M. Tam, "Multi-level optimal design of buildings with active control under winds using genetic algorithms," Journal of Wind Engineering and Industrial Aerodynamics, vol. 86, no. 1, pp. 65–86, 2000. · ·

60. H. Li, Z. Chen, and H. Polinder, "Optimization of multibrid permanent-magnet wind generator systems,"IEEE Transactions on Energy Conversion, vol. 24, no. 1, pp. 82–92, 2009. · ·

61. S. A. Grady, M. Y. Hussaini, and M. M. Abdullah, "Placement of wind turbines using genetic algorithms," Renewable Energy, vol. 30, no. 2, pp. 259–270, 2005. · ·

62. A. Emami and P. Noghreh, "New approach on optimization in placement of wind turbines within wind farm by genetic algorithms," Renewable Energy, vol. 35, no. 7, pp. 1559–1564, 2010. ··

63. Varun and Siddhartha, "Thermal performance optimization of a flat plate solar air heater using genetic algorithm," Applied Energy, vol. 87, no. 5, pp. 1793–1799, 2010. ··

64. M. Zagrouba, A. Sellami, M. Bouaïcha, and M. Ksouri, "Identification of PV solar cells and modules parameters using the genetic algorithms: application to maximum power extraction," Solar Energy, vol. 84, no. 5, pp. 860–866, 2010. ··

65. K. Tselepidou and K. L. Katsifarakis, "Optimization of the exploitation system of a low enthalpy geothermal aquifer with zones of different transmissivities and temperatures," Renewable Energy, vol. 35, no. 7, pp. 1408–1413, 2010. ··

66. S. H. El-Hefnawi, "Photovoltaic diesel-generator hybrid power system sizing," Renewable Energy, vol. 13, no. 1, pp. 33–40, 1998.

67. R. Dufo-López and J. L. Bernal-Agustín, "Design and control strategies of PV-diesel systems using genetic algorithms," Solar Energy, vol. 79, no. 1, pp. 33–46, 2005.

68. E. Koutroulis, D. Kolokotsa, A. Potirakis, and K. Kalaitzakis, "Methodology for optimal sizing of stand-alone photovoltaic/wind-generator systems using genetic algorithms," Solar Energy, vol. 80, no. 9, pp. 1072–1088, 2006. ··

69. R. Dufo-López, J. L. Bernal-Agustín, and J. Contreras, "Optimization of control strategies for stand-alone renewable energy systems with hydrogen storage," Renewable Energy, vol. 32, no. 7, pp. 1102–1126, 2007.

70. T. Senjyu, D. Hayashi, A. Yona, N. Urasaki, and T. Funabashi, "Optimal configuration of power generating systems in isolated island with renewable energy," Renewable Energy, vol. 32, no. 11, pp. 1917–1933, 2007. ··

71. D. Fogel and K. Chellapilla, "Revisiting evolutionary programming," in Applications and Science of Computational Intelligence, Proceedings of SPIE, pp. 2–11, Orlando, Fla, USA, 1998. ·

72. T. D. H. Cau and R. J. Kaye, "Multiple distributed energy storage scheduling using constructive evolutionary programming," in Proceedings of the 22nd IEEE International Conference on Power Industry Computer Applications, pp. 402–407, May 2001.

73. K. F. Fong, V. I. Hanby, and T. T. Chow, "HVAC system optimization for energy management by evolutionary programming," Energy and Buildings, vol. 38, no. 3, pp. 220–231, 2006. ··

74. I. F. MacGill, "An evolutionary programming tool for assessing the operational value of distributed energy resources within restructured electricity industries," in Proceedings of the Australasian Universities Power Engineering (AUPEC ‹07), pp. 1–6, Australasian Universities, December 2007. ··

75. S. Chen, Y. Zheng, C. Cattani, and W. Wang, "Modeling of biological intelligence for SCM system optimization," Computational and Mathematical Methods in Medicine, vol. 2012, Article ID 769702, 10 pages, 2012. ···

76. Y. C. Chang, "Optimal chiller loading by evolution strategy for saving energy," Energy and Buildings, vol. 39, no. 4, pp. 437–444, 2007. ··

77. T. Logenthiran, D. Srinivasan, A. M. Khambadkone, and T. Sundar Raj, "Optimal sizing of distributed energy resources for integrated microgrids using evolutionary strategy," in Proceedings of IEEE Congress on Evolutionary Computation, pp. 1–8, 2012.

78. M. A. Falcone, H. S. Lopes, and L. dos Santos Coelho, "Supply chain optimisation using evolutionary algorithms," International Journal of Computer Applications in Technology, vol. 31, no. 3-4, pp. 158–167, 2008. ··

79. S. Chakraborty, T. Senjyu, A. Yona, A. Y. Saber, and T. Funabashi, "Fuzzy unit commitment strategy integrated with solar energy system using a modified differential evolution approach," in Proceedings of Asia and Pacific Conference & Exposition on Transmission & Distribution, pp. 1–4, October 2009. ··

80. L. Slimani and T. Bouktir, "Application of differential evolution algorithm to optimal power flow with high wind energy penetration," Acta Electrotehnica, vol. 53, no. 1, pp. 59–68, 2012.

81. L. dos Santos Coelho, A. D. V. De Almeida, and V. C. Mariani, "Cultural differential evolution approach to optimize the economic dispatch of electrical energy using thermal generators," in Proceedings of the 13th IEEE International Conference on Emerging Technologies and Factory Automation (ETFA ‹08), pp. 1378–1383, September 2008. · ·

82. R. Suzuki, F. Kawai, S. Kitagawa et al., "The ε constrained differential evolution approach for optimal operational planning of energy plants," in Proceedings of IEEE Congress on Evolutionary Computation (CEC ‹10), July 2010. · ·

83. H. A. Hejazi, H. R. Mohabati, S. H. Hosseinian, and M. Abedi, "Differential evolution algorithm for security-constrained energy and reserve optimization considering credible contingencies," IEEE Transactions on Power Systems, vol. 26, no. 3, pp. 1145–1155, 2011. · ·

84. W. S. Lee, Y. T. Chen, and Y. Kao, "Optimal chiller loading by differential evolution algorithm for reducing energy consumption," Energy and Buildings, vol. 43, no. 2-3, pp. 599–604, 2011. · ·

85. L. Peng, Y. Wang, G. Dai, Y. Chang, and F. Chen, "Optimization of the Earth-Moon low energy transfer with differential evolution based on uniform design," in Proceedings of IEEE Congress on Evolutionary Computation (CEC ‹10), July 2010. · ·

86. N. Srinivas and K. Deb, "Multiobjective optimization using nondominated sorting in genetic algorithms,"Evolutionary Computing, vol. 2, pp. 221–248, 1994.

87. K. Deb, A. Pratap, S. Agarwal, and T. Meyarivan, "A fast and elitist multiobjective genetic algorithm: NSGA-II," IEEE Transactions on Evolutionary Computation, vol. 6, no. 2, pp. 182–197, 2002. · ·

88. E. Zitzler and L. Thiele, "Multiobjective evolutionary algorithms: a comparative case study and the strength Pareto approach," IEEE Transactions on Evolutionary Computation, vol. 3, no. 4, pp. 257–271, 1999.

89. E. Zitzler, M. Laumanns, and L. Thiele, "SPEA2: improving the strength Pareto evolutionary algorithm," in Evolutionary Methods for Design, Optimization and Control with Applications to Industrial Problems, K. Giannakoglou, D. Tsahalis, J. Periaux, P. Papailou, and T. Fogarty, Eds., Athens, Greece, 2001.

90. J. D. Knowles and D. W. Corne, "M-PAES: a memetic algorithm for multiobjective optimization," inProceedings of the Congress on Evolutionary Computation (CEC ‹00), vol. 1, pp. 325–332, July 2000.

91. H. A. Abbass, R. Sarker, and C. Newton, "PDE: a pareto-frontier differential evolution approach for multi-objective optimization problems," in Proceedings of IEEE Congress on Evolutionary Computation, vol. 2, pp. 971–978, May 2001.

92. R. Angira and B. V. Babu, "Non-dominated sorting differential evolution (NSDE): an extension of differential evolution for multi-objective optimization," in Proceedings of 2nd Indian International Conference on Artificial Intelligence, 2005.

93. Y. J. Zheng, Q. Song, and S. Y. Chen, "Multiobjective fireworks optimization for variable-rate fertilization in oil crop production," Applied Soft Computing. In press.

94. E. Benini and A. Toffolo, "Optimal design of horizontal-axis wind turbines using blade-element theory and evolutionary computation," Journal of Solar Energy Engineering, vol. 124, no. 4, pp. 357–363, 2002. · ·

95. M. Zhao, Z. Chen, and F. Blaabjerg, "Optimisation of electrical system for offshore wind farms via genetic algorithm," IET Renewable Power Generation, vol. 3, no. 2, pp. 205–216, 2009. · ·

96. A. Kusiak, Z. Zhang, and M. Li, "Optimization of wind turbine performance with data-driven models,"IEEE Transactions on Sustainable Energy, vol. 1, no. 2, pp. 66–76, 2010. · ·

97. A. Kusiak and Z. Song, "Design of wind farm layout for maximum wind energy capture," Renewable Energy, vol. 35, no. 3, pp. 685–694, 2010. · ·

98. J. L. Bernal-Agustín, R. Dufo-López, and D. M. Rivas-Ascaso, "Design of isolated hybrid systems minimizing costs and pollutant emissions," Renewable Energy, vol. 31, no. 14, pp. 2227–2244, 2006. · ·

99. R. Dufo-López and J. L. Bernal-Agustín, "Multi-objective design of PV-wind-diesel-hydrogen-battery systems," Renewable Energy, vol. 33, no. 12, pp. 2559–2572, 2008. · ·

100. B. Ould Bilal, V. Sambou, P. A. Ndiaye, C. M. F. Kébé, and M. Ndongo, "Optimal design of a hybrid solar-wind-battery system using the minimization of the annualized cost system and the minimization of the loss of power supply probability (LPSP)," Renewable Energy, vol. 35, no. 10, pp. 2388–2390, 2010. · ·

101. F. G. Montoya, R. Banos, C. Gil, A. Espin, A. Alcayde, and J. Gomez, "Minimization of voltage deviation and power losses in power networks using Pareto optimization methods," Engineering Applications of Artificial Intelligence, vol. 23, pp. 695–703, 2010.

102. Y. Thiaux, J. Seigneurbieux, B. Multon, and H. Ben Ahmed, "Load profile impact on the gross energy requirement of stand-alone photovoltaic systems," Renewable Energy, vol. 35, no. 3, pp. 602–613, 2010. · ·

103. P. Rao and C. H. Peng, "A research on power dispatch of energy-saving and emission-reduction generation based on the improved differential evolution algorithm," Journal of East China Jiaotong University, vol. 5, pp. 48–52, 2010.

104. P. Tarasewich and P. R. McMullen, "Swarm intelligence powers in numbers," Communications of the ACM, vol. 45, no. 8, pp. 62–67, 2002. · ·

105. E. Bonabeau, M. Dorigo, and G. Theraulaz, Swarm Intelligence: From Natural to Artificial Systems, Oxford University Press, New York, NY, USA, 1999.

106. J. Kennedy and R. Eberhart, "Particle swarm optimization," in Proceedings of the IEEE International Conference on Neural Networks, pp. 1942–1948, Perth, Australia, December 1995.

107. A. Colorni, M. Dorigo, and V. Maniezzo, "Distributed optimization by ant colonies," in Proceedings of European Conference on Artificial Life, pp. 134–142, Paris, France, 1991.

108. X. S. Yang, "Engineering optimizations via nature-inspired virtual bee algorithms," in Proceedings of the 1st International Work-Conference on the Interplay Between Natural and Artificial Computation (IWINAC ‹05), vol. 3562 of Lecture Notes in Computer Science, pp. 317–323, Springer, Las Palmas, Spain, June 2005.

109. D. Karaboga and B. Basturk, "A powerful and efficient algorithm for numerical function optimization: artificial bee colony (ABC) algorithm," Journal of Global Optimization, vol. 39, no. 3, pp. 459–471, 2007. · · · ·

110. J. D. Farmer, N. H. Packard, and A. S. Perelson, "The immune system, adaptation, and machine learning," Physica D, vol. 22, no. 1–3, pp. 187–204, 1986. · · ·

111. J. Kennedy, "Bare bones particle swarms," in Proceedings of IEEE Swarm Intelligence Symposium, pp. 120–127, IEEE Press, 2003.

112. M. R. AlRashidi and K. M. EL-Naggar, "Long term electric load forecasting based on particle swarm optimization," Applied Energy, vol. 87, no. 1, pp. 320–326, 2010. · ·

113. T. Niknam and B. B. Firouzi, "A practical algorithm for distribution state estimation including renewable energy sources," Renewable Energy, vol. 34, no. 11, pp. 2309–2316, 2009. · ·

114. N. Amjady and H. R. Soleymanpour, "Daily Hydrothermal Generation Scheduling by a new Modified Adaptive Particle Swarm Optimization technique," Electric Power Systems Research, vol. 80, no. 6, pp. 723–732, 2010. · ·

115. Z. H. Zhan, J. Zhang, Y. Li, and H. S. H. Chung, "Adaptive particle swarm optimization," IEEE Transactions on Systems, Man, and Cybernetics, Part B, vol. 39, no. 6, pp. 1362–1381, 2009. · ·

116. Y.-J. Zheng, H.-F. Ling, and Q. Guan, "Adaptive parameters for a modified comprehensive learning particle swarm optimizer," Mathematical Problems in Engineering, vol. 2013, 11 pages, 2013. ·

117. T. Y. Lee, "Short term hydroelectric power system scheduling with wind turbine generators using the multi-pass iteration particle swarm optimization approach," Energy Conversion and Management, vol. 49, no. 4, pp. 751–760, 2008. · ·

118. C. Kongnam and S. Nuchprayoon, "A particle swarm optimization for wind energy control problem," Renewable Energy, vol. 35, no. 11, pp. 2431–2438, 2010. · ·

119. S. Khanmohammadi, M. Amiri, and M. T. Haque, "A new three-stage method for solving unit commitment problem," Energy, vol. 35, no. 7, pp. 3072–3080, 2010. · ·

120. P. R. López, F. Jurado, N. Ruiz-Reyes, S. García Galán, and M. Gómez, "Particle swarm optimization for biomass-fuelled systems with technical constraints," Engineering Applications of Artificial Intelligence, vol. 21, no. 8, pp. 1389–1396, 2008. · ·

121. P. R. Lopez, S. G. Galan, N. Ruiz-Reyes, and F. Jurado, "A method for particle swarm optimization and its application in location of biomass power plants," International Journal of Green Energy, vol. 5, no. 3, pp. 199–211, 2008. · ·

122. T. Y. Lee and C. L. Chen, "Wind-photovoltaic capacity coordination for a time-of-use rate industrial user," IET Renewable Power Generation, vol. 3, no. 2, pp. 152–167, 2009. · ·

123. A. K. Kaviani, G. H. Riahy, and S. M. Kouhsari, "Optimal design of a reliable hydrogen-based stand-alone wind/PV generating system, considering component outages," Renewable Energy, vol. 34, no. 11, pp. 2380–2390, 2009. · ·

124. A. Kornelakis and E. Koutroulis, "Methodology for the design optimisation and the economic analysis of grid-connected photovoltaic systems," IET Renewable Power Generation, vol. 3, no. 4, pp. 476–492, 2009. · ·

125. A. Kornelakis and Y. Marinakis, "Contribution for optimal sizing of grid-connected PV-systems using PSO," Renewable Energy, vol. 35, no. 6, pp. 1333–1341, 2010. · ·

126. S. M. Hakimi and S. M. Moghaddas-Tafreshi, "Optimal sizing of a stand-alone hybrid power system via particle swarm optimization for Kahnouj area in south-east of Iran," Renewable Energy, vol. 34, no. 7, pp. 1855–1862, 2009. · ·

127. A. Mahor, V. Prasad, and S. Rangnekar, "Economic dispatch using particle swarm optimization: a review," Renewable and Sustainable Energy Reviews, vol. 13, no. 8, pp. 2134–2141, 2009. · ·

128. Z. L. Gaing, "Particle swarm optimization to solving the economic dispatch considering the generator constraints," IEEE Transactions on Power Systems, vol. 18, no. 3, pp. 1187–1195, 2003. · ·

129. D. N. Jeyakumar, T. Jayabarathi, and T. Raghunathan, "Particle swarm optimization for various types of economic dispatch problems," International Journal of Electrical Power & Energy Systems, vol. 28, no. 1, pp. 36–42, 2006. · ·

130. L. Wang and C. Singh, "Reserve-constrained multiarea environmental/economic dispatch based on particle swarm optimization with local search," Engineering Applications of Artificial Intelligence, vol. 22, no. 2, pp. 298–307, 2009. · ·

131. K. K. Mandal, M. Basu, and N. Chakraborty, "Particle swarm optimization technique based short-term hydrothermal scheduling," Applied Soft Computing Journal, vol. 8, no. 4, pp. 1392–1399, 2008. · ·

132. R. Chakrabarti, P. K. Chattopadhyay, M. Basu, and C. K. Panigrahi, "Particle swarm optimization technique for dynamic economic dispatch," Journal of the Institution of Engineers, vol. 87, pp. 48–54, 2006.

133. J. B. Park, Y. W. Jeong, H. H. Kim, and J. R. Shin, "An improved particle swarm optimization for economic load dispatch with valve point effect," International Journal of Innovations in Energy Systems and Power, vol. 1, no. 1, 2006.

134. L. dos Santos Coelho and V. C. Mariani, "Economic dispatch optimization using hybrid chaotic particle swarm optimizer," in Proceedings of IEEE International Conference on Systems, Man, and Cybernetics (SMC ‹07), pp. 1963–1968, October 2007. · ·

135. J. Cai, X. Ma, L. Li, and P. Haipeng, "Chaotic particle swarm optimization for economic dispatch considering the generator constraints," Energy Conversion and Management, vol. 48, no. 2, pp. 645–653, 2007. · ·

136. A. I. S. Kumar, K. Dhanushkodi, J. J. Kumar, and C. K. C. Paul, "Particle swarm optimization solution to emission and economic dispatch problem," in Proceedings of IEEE Confernce on Covergent Technologies for the Asia-Pacific Region (TENCON ‹03), vol. 1, pp. 435–439, October 2003.

137. B. Zhao, C. Guo, and Y. Cao, "Dynamic economic dispatch in electricity market using particle swarm optimization algorithm," in Proceedings of 5th World Congress on Intelligent Control and Automation (WCICA ‹04), pp. 5050–5054, June 2004.

138. M. A. Abido, "Multiobjective particle swarm for environmental/economic dispatch problem," inProceedings of the 8th International Power Engineering Conference (IPEC ‹07), pp. 1385–1390, December 2007.

139. M. A. Alrashidi and M. E. Hawary, "Impact of loading conditions on the emission economic dispatch," in Proceedings of World Academy of Science and Engineering and Technology, pp. 148–151, 2008.

140. Z. Li, X. P. Chang, and J. H. Qin, "Application of ant colony algorithms to optimization design of solar energy dynamic power system in space station," Proceedings of the Chinese Society of Electrical Engineering, vol. 25, pp. 294–298, 2005.

141. D. Xu, L. Kang, and B. Cao, "Graph-based ant system for optimal sizing of standalone hybrid wind/PV power systems," in Computational Intelligence, vol. 4114 of Lecture Notes in Computer Science, pp. 1136–1146, Springer, 2006.

142. W. K. Foong, H. R. Maier, and A. R. Simpson, "Power plant maintenance scheduling using ant colony optimization: an improved formulation," Engineering Optimization, vol. 40, no. 4, pp. 309–329, 2008. · · ·

143. L. Warner and U. Vogel, "Optimization of energy supply networks using ant colony optimization," inProceedings of 22th International Conference on Informatics for Environmental Protection, pp. 327–334, 2008.

144. P. C. See, V. C. Tai, and M. Molinas, "Ant colony optimization applied to control of ocean wave energy converters," Energy Procedia, vol. 20, pp. 148–155, 2012.

145. M. D. Toksari, "Estimating the net electricity energy generation and demand using the ant colony optimization approach: case of Turkey," Energy Policy, vol. 37, no. 3, pp. 1181–1187, 2009. · ·

146. O. Baskan, S. Haldenbilen, H. Ceylan, and H. Ceylan, "Estimating transport energy demand using ant colony optimization," Energy Sources, Part B, vol. 7, no. 2, pp. 188–199, 2012. ·

147. A. Afshar, O. Bozorg Haddad, M. A. Mariño, and B. J. Adams, "Honey-bee mating optimization (HBMO) algorithm for optimal reservoir operation," Journal of the Franklin Institute, vol. 344, no. 5, pp. 452–462, 2007. · ·

148. T. Niknam, S. I. Taheri, J. Aghaei, S. Tabatabaei, and M. Nayeripour, "A modified honey bee mating optimization algorithm for multiobjective placement of renewable energy resources," Applied Energy, vol. 88, no. 12, pp. 4817–4830, 2011.

149. T. Niknam, H. D. Mojarrad, H. Z. Meymand, and B. B. Firouzi, "A new honey bee mating optimization algorithm for non-smooth economic dispatch," Energy, vol. 36, no. 2, pp. 896–908, 2011. · ·

150. F. S. Abu-Mouti and M. E. El-Hawary, "Optimal dynamic economic dispatch including renewable energy source using artificial bee colony algorithm," in Proceedings of IEEE Systems Conference, 2012.

151. D. Vera, J. Carabias, F. Jurado, and N. Ruiz-Reyes, "A Honey Bee Foraging approach for optimal location of a biomass power plant," Applied Energy, vol. 87, no. 7, pp. 2119–2127, 2010. · ·

152. W. C. Hong, "Electric load forecasting by seasonal recurrent SVR (support vector regression) with chaotic artificial bee colony algorithm," Energy, vol. 36, no. 9, pp. 5568–5578, 2011. ·

153. T. K. Abdul Rahman, Z. M. Yasin, and W. N. W. Abdullah, "Artificial-immune-based for solving economic dispatch in power system," in Proceedings of National Power and Energy Conference (PECon ‹04), pp. 31–35, November 2004.

154. L. D. S. Coelho and V. C. Mariani, "Chaotic artificial immune approach applied to economic dispatch of electric energy using thermal units," Chaos, Solitons & Fractals, vol. 40, no. 5, pp. 2376–2383, 2009. · ·

155. P. Arsalani and M. Seddighizadeh, "Minimizing the loss of energy in transmission systems with capacitor placement using an immune algorithm and fuzzy logic," in Proceedings of 2nd Conference on Energy Management and Conservation, 2012.

156. A. Mellit and S. A. Kalogirou, "Application of neural networks and genetic algorithms for predicting the optimal sizing coefficient of photovoltaic supply (PVS) systems," in Proceedings of the World Renewable Energy Congress IX and Exhibition, 2006.

157. A. Mellit, "ANFIS-based genetic algorithm for predicting the optimal sizing coefficient of photovoltaic supply (PVS) systems," in Proceedings of 3rd International Conference on Thermal Engineering: Theory and Applications, pp. 96–102, 2007.

158. Y. P. Chang and C. N. Ko, "A PSO method with nonlinear time-varying evolution based on neural network for design of optimal harmonic filters," Expert Systems with Applications, vol. 36, no. 3, pp. 6809–6816, 2009. · ·

159. A. Li, L. Wang, J. Li, and C. Ji, "Application of immune algorithm-based particle swarm optimization for optimized load distribution among cascade hydropower stations," Computers and Mathematics with Applications, vol. 57, no. 11-12, pp. 1785–1791, 2009. · ·

160. Z. X. Yang, X. F. Yue, and L. Wang, "Study on the energy consumption optimization in a central air-conditioning system based on bee evolutionary genetic algorithm method," Building Science, vol. 27, no. 6, pp. 78–82, 2011.

161. M. S. Kıran, E. Özceylan, M. Gündüz, and T. Paksoy, "A novel hybrid approach based on particle swarm optimization and ant colony algorithm to forecast energy demand of Turkey," Energy Conversion and Management, vol. 53, no. 1, pp. 75–83, 2012.

162. A. Ghanbari, M. Kazemi, F. Mehmanpazir, and M. M. Nakhostin, "A cooperative ant colony optimization-genetic algorithm approach for construction of energy demand forecasting knowledge-based expert systems," Knowledge-Based Systems, vol. 39, pp. 194–206, 2013. ·

163. B. Tudu, S. Majumder, K. K. Mandal, and N. Chakraborty, "Comparative performance study of genetic algorithm and particle swarm optimization applied on off-grid renewable hybrid energy system," in Swarm, Evolutionary, and Memetic Computing, vol. 7076 of Lecture Notes in Computer Science, pp. 151–158, Springer, 2011. ·

164. M. Carlini and S. Castellucci, "Modelling and simulation for energy production parametric dependence in greenhouses," Mathematical Problems in Engineering, vol. 2010, Article ID 590943, 28 pages, 2010. · ·

165. M. Carlini, S. Castellucci, M. Guerrieri, and T. Honorati, "Stability and control for energy production parametric dependence," Mathematical Problems in Engineering, vol. 2010, Article ID 842380, 21 pages, 2010. · ·

166. M. Fadaee and A. M. Radzi, "Multi-objective optimization of a stand-alone hybrid renewable energy system by using evolutionary algorithms: a review," Renewable and Sustainable Energy Reviews, vol. 16, no. 5, pp. 3364–3369, 2012.

167. J. Gruska, Quantum Computing, Advanced Topics in Computer Science Series, McGraw-Hill, New York, NY, USA, 1999.

168. G. Păun, G. Rozenberg, and A. Salomaa, DNA Computing: New Computing Paradigms, Texts in Theoretical Computer Science. An EATCS Series, Springer, Berlin, Germany, 1998.

169. M. Joyeux, S. Buyukdagli, and M. Sanrey, "1/f fluctuations of DNA temperature at thermal denaturation,"Physical Review E, vol. 75, no. 6, Article ID 061914, 9 pages, 2007. · ·

170. A. Castro, M. A. L. Marques, D. Varsano, F. Sottile, and A. Rubio, "The challenge of predicting optical properties of biomolecules: what can we learn from time-dependent density-functional theory?" Comptes Rendus Physique, vol. 10, no. 6, pp. 469–490, 2009. · ·

171. C. Cattani, E. Laserra, and I. Bochicchio, "Simplicial approach to fractal structures," Mathematical Problems in Engineering, vol. 2012, Article ID 958101, 21 pages, 2012. · ·

172. C. Cattani, "On the existence of wavelet symmetries in archaea DNA," Computational and Mathematical Methods in Medicine, Article ID 673934, 21 pages, 2012. · · ·

173. G. Zhang, "Quantum-inspired evolutionary algorithms: a survey and empirical study," Journal of Heuristics, vol. 17, no. 3, pp. 303–351, 2011. · ·

174. E. G. Bakhoum and C. Toma, "Dynamical aspects of macroscopic and quantum transitions due to coherence function and time series events," Mathematical Problems in Engineering, vol. 2010, Article ID 428903, 13 pages, 2010. · ·

175. E. G. Bakhoum and C. Toma, "Mathematical transform of traveling-wave equations and phase aspects of quantum interaction," Mathematical Problems in Engineering, vol. 2010, Article ID 695208, 15 pages, 2010.

Chapter 5

Greenhouse Gas Emission Accounting and Management of Low-Carbon Community

Dan Song, Meirong Su, Jin Yang, and Bin Chen

State Key Laboratory of Water Environment Simulation, School of Environment, Beijing Normal University, Beijing 100875, China

ABSTRACT

As the major source of greenhouse gas (GHG) emission, cities have been under tremendous pressure of energy conservation and emission reduction for decades. Community is the main unit of urban housing, public facilities, transportation, and other properties of city's land use. The construction of low-carbon community is an important pathway to realize carbon emission mitigation in the context of rapid urbanization. Therefore, an efficient carbon accounting framework should be proposed for CO_2 emissions mitigation at a subcity level. Based on life-cycle analysis (LCA), a three-tier accounting framework for the carbon emissions of the community is put forward, including emissions from direct fossil fuel combustion, purchased energy (electricity, heat, and water), and supply chain emissions embodied in the consumption of goods. By compiling a detailed CO_2 emission inventory, the magnitude of carbon emissions and the mitigation potential in a typical high-quality community in Beijing are quantified within the accounting framework proposed. Results show that emissions from supply chain emissions embodied in the consumption of goods cannot be ignored. Specific suggestions are also provided for the urban decision makers to achieve the optimal resource allocation and further promotion of low-carbon communities.

INTRODUCTION

Global warming has been a hot topic since a few decades ago and became a direct trigger for behavior change for people worldwide [1–35]. As the most impacted region by human activities, cities emit more than 75% of the total greenhouse gas, in which CO_2 occupies a large proportion [36]. Cities play an important role in global carbon cycle, and most of their impacts are exerted via indirect pathways [37]. With the purpose of the energy resource consumption minimization and greenhouse gas emission reduction, low-carbon cities have attracted increasing attention [38]. As the cell of a city, community is the basic unit in the low-carbon city construction, and its structure and density also play a key role in energy consumption and CO_2 emission [39,40]. Low-carbon community provides a platform for individual behavior change [41, 42]. The UK Low-Carbon Transition Plan [43] also makes explicit the major role that households and communities play in building a low-carbon future. A common viewpoint has been reached that low-carbon community will be an efficient way to achieve sustainable development due to its energy utilization, internal structure optimization, and external effects reduction. Obviously, the pursuit of low-carbon community would be extremely essential to retard the global climate change.

In order to estimate the contribution of cities to global climate change, many attempts have been made to quantify the carbon emissions associated with the accounting level in the community. Recently, many organizations have been conducting "low-carbon" projects to estimate the contributions to global climate change. Many protocols were put out to guide organizations to measure GHG emissions [44–46]. These protocols are mainly concentrated on direct emissions and indirect emissions from purchased energy, with less focus on supply chain emissions that occupied a large proportion in a community. For example, direct CO_2 emissions are found to be generated by direct household energy use, whereas indirect CO_2 emissions are generated in the industrial sectors producing nonenergy commodities demanded by the households [47]. Pachauri and Spreng applied the IO models into the calculation of direct and indirect energy consumption of households in India based on the 115-sector classification input-output tables [48]. Lu et al. quantified the direct and indirect household emissions of CO_2 in China with the help of input-output life-cycle assessment (IO-LCA) combined with 8 categories of household expenditure [49]. A calculation framework for whole life-circle carbon budget in residential area was presented based on building system, social system, and green space system, showing that the ratio of carbon source to carbon sink is 29:1 and that of society source to building source is 4.6:1 [50]. It can be seen that there is serious imbalance between carbon sink and carbon source in this residential area, and the society source is a key factor for carbon budget balance.

Moreover, Matthews et al. classified the variety scopes of carbon footprint into 3 tiers, including direct emissions, emissions from purchased energy, and supply chain emissions [51]. In their study, two case studies of book publishers and power generation were conducted, which illustrated that the first 2-tier emissions accounted for only a small part while a large portion is constituted by emissions embodied in the supply chain. The Scope 3 footprints of US economic sectors using a modified form of the 2002 US benchmark Economic Input-Output Life Cycle Assessment (EIO-LCA) model was developed to categorize upstream emission sources [52]. Larsen and Hertwich developed a greenhouse gas emissions inventory related to the provision of municipal services in the city of Trondheim, Norway, indicating that approximately 93% of the total carbon footprint of municipal services is indirect emissions [53]. The authors also established CO_2 inventories focused on the supply chain emissions of CO_2 emissions from each sector, for example, agriculture, industry, transportation, and tertiary industry, and identified the sectors that contribute the most to climate change [54].

As can be seen, the previous studies on 3-tier accounting are mainly concentrated on industry sectors, with less focus on community-level CO_2 emissions. A special focus should be transferred to identify Scope 3 categories that are relevant and incorporated into the footprint analysis. Thus, further characterization of the total supply chain emissions in community is necessary in order to achieve a better strategy for carbon emission mitigation. Approaches based on life cycle assessment (LCA) methods are available to estimate the embodied CO_2 in the consumption of goods, which provides a framework for analysis of the potential environmental impacts embodied throughout the lifetime of goods [55, 56]. There are two common types of LCA models, that is, process-based LCA and EIO-LCA, varying according to differences in system scope and analysis with its own processes and characteristics [57]. Economic IO models were first developed by Leontief in 1936 to aid manufacturing planning [58]. Compared to the process-based LCA, EIO-LCA addresses some of the drawbacks of process-based LCA model and greatly expands the system scope to include the entire economy of a region, which can assess the energy consumption and environmental impacts of goods from a nationwide perspective based on economic input-output matrix.

The aim of this paper is to propose an efficient three-tier carbon emission accounting framework for community. Taking a typical high-quality community in Beijing as case study, this study also intends to quantify the magnitude of carbon emissions and the mitigation potential using the method of LCA in combination with a detailed CO_2 emission inventory, including emissions from direct fossil fuel combustion, emissions from purchased energy (mainly contains electricity, water, and heat), and supply chain emissions embodied in the consumption of goods. Some suggestions about the realization of optimized resource allocation and further promotion of such communities are also given for the decision makers.

METHODOLOGY

We develop estimation equations for three tiers of carbon footprint of the community based on the scope initially developed by Matthews et al. [51].

Tier 1 includes direct emissions from household fossil fuel combustion and vehicles, including emissions from natural gas, gasoline, diesel oil, and jet kerosene. This is similar to the "consumer perspective" used for emissions inventories [59].

Tier 2 is based on Tier 1, in addition to indirect emissions from purchased energy (mainly contains electricity, water, and heat) for a community.

Tier 3 includes the total supply chain emissions embodied in the consumption of goods and activities. The accounting model and boundaries used for estimating all purchases and activities aspects in a supply chain by any sector of a community are based on EIO-LCA, which are consistent with the data structure described in Section 3.2.

The decomposition analysis is carried out in two steps. Firstly, Tier 1 and Tier 2 CO_2 emissions from household energy use are analyzed using a simple energy emission model. Secondly, Tier 3 CO_2 emissions are analyzed using an extended LCA model that also incorporates energy and emission matrices.

In terms of spatial system boundary, the total CO_2 emissions are derived from emissions from household and public area. Thus the total CO_2 emissions calculated in 3 tiers can be defined as

$$E = E_h + E_p,$$

$$E_h = E_{h1} + E_{h2} + E_{h3} + E_{h4} + E_{h5} + E_{h6} + E_{h7} + E_{h8} + E_{h9},$$

$$E_p = E_{p1} + E_{p2}, \tag{1}$$

where E is the total CO_2 emissions from community; E_h refers to all the three-tiers CO_2 emissions from household that consists of CO_2 emissions from direct energy consumption (E_{h1}), indirect energy and water consumption (E_{h2}), transport and community (E_{h3}), food (E_{h4}), clothing and footwear (E_{h5}), household appliances and services (E_{h6}), healthcare (E_{h7}), education and recreation (E_{h8}), and from buildings (E_{h9}); Ep refers to CO_2 emissions from the public area of a community that consisted of CO_2 emissions from electricity consumption (E_{p1}) and from water consumption (E_{p2}).

CASE STUDY

Study Area

As Beijing is in its fast process of urbanization, community construction turns into a key element of the city renovation. This paper selects a typical high-quality community in Beijing as the case study. The community covers an area of $8.2 \times 10^3 \, m^2$ with a construction area of $3.0 \times 10^5 \, m^2$ and a living area of $9.0 \times 10^4 \, m^2$. The community has 1630 households and a permanent population of 3100, with a green space of more than $2500 \, m^2$ and a greening rate of 30%. The community has carried out the garbage classification since 2004. So far, the capacity of the kitchen waste disposal equipment that came into use has reached 20 kg per day. The power consumption is $2.24 \times 10^5 \, kWh$ per month, and water consumption is about $1.63 \times 10^4 \, m^3$ per month.

Data Analysis

CO_2 emission factors of primary energy are based on the CO_2 content of the fuels and the type of energy, which are elaborated in IPCC [60]. CO_2 emissions factors of electricity are based on coal factors but corrected by standard coal consumption of power supply (standard coal consumption 356 g/kWh, the average value in China [61]). CO_2 emissions factors for renewable energy are considered to be zero. The CO_2 emissions factors of energy shown in Table 1. Other CO_2 emission factors of consumption goods can be referred to the embodied greenhouse gas emission database [62].

Table 1: The CO_2 emissions factors of conventional energy

	Coal ($KgCO_2$/GJ)	Natural gas ($KgCO_2$/GJ)	Electricity ($KgCO_2$/kWh)	Gasoline ($KgCO_2$/GJ)
CO_2 emission coefficients	110.08	56.10	1.15	69.30

In this study, direct CO_2 emissions from the consumption of electricity and heating are not considered. The energy inputs for the production of electricity and district heating are estimated as the final consumption of energy production; that is, all emissions caused by energy production are specified for each of the fuel inputs [56].

The consumption data are developed based on the survey carried out in the community. Based on the previous studies engaged to classify the sectoral composition of consumption [48, 49, 63], we aggregate the community consumption in the database into the same expenditure framework, of which 8 emission categories include food, clothing and footwear, household appliances and services, health care, transport and communication, education and recreation, building, and miscellaneous goods, as listed in Table 2.

Table 2: Consumption categories of the community

No.	Items	Contents
1	Food	Miscellaneous food products, beverages, and tobacco products.
2	Clothing and footwear	Miscellaneous textile products, leather footwear.
3	Household appliances and services	Electrical appliances (television, computer, and other electrical machinery). Furniture and fixtures, wood products, and kitchen appliances.
4	Healthcare	Cosmetics, medical and health services, and other services.
5	Transport and communication	Communication equipments, ships and boats, railway, motor vehicles, bicycles, other transportation ways, and other transport services.
6	Education and recreation	Paper, paper products and newspapers, printing publishing and similar activities, and education and research.
7	Buildings	Residence and public buildings.
8	Misc goods and service	Trade, banking, insurance, and so forth.
9	Direct energy consumption	Natural gas, gasoline, diesel oil, and jet kerosene.
10	Indirect energy and water consumption	Electricity, heat, and water.

RESULTS

Comparison of Tier 1, Tier 2, and Tier 3 CO_2 Emissions

The results show that the first 2 tiers defined by the current most carbon footprint protocols only occupy a small fraction of the total supply chain (Tier 3). Direct emissions from the community are only 1.58% of the total emissions, and on average only 11.46% of Tier3 are captured by Tier 2. The major carbon source is the total supply chain emissions embodied in the consumption of goods and activities, which is called Tier 3. Thus reduction emphasis should be put on Tier 3. From this aspect we can see that a large quantity of CO_2 emissions may be underestimated according to the current estimation protocols.

CO_2 Emissions Structure

For the total CO_2 emissions, which are defined as Tier 3, the top 3 emission items are transport and communication (41.36%), buildings (14.11%), and education and recreation (10.41%), as shown in Figure 1. Income is an important factor for CO_2 emission. In a typical high-quality community of Beijing, residents enjoy a high-standard life and prefer more convenient and faster communication tools. Thus more private cars and advanced communication tools are needed, which add to the total emissions.

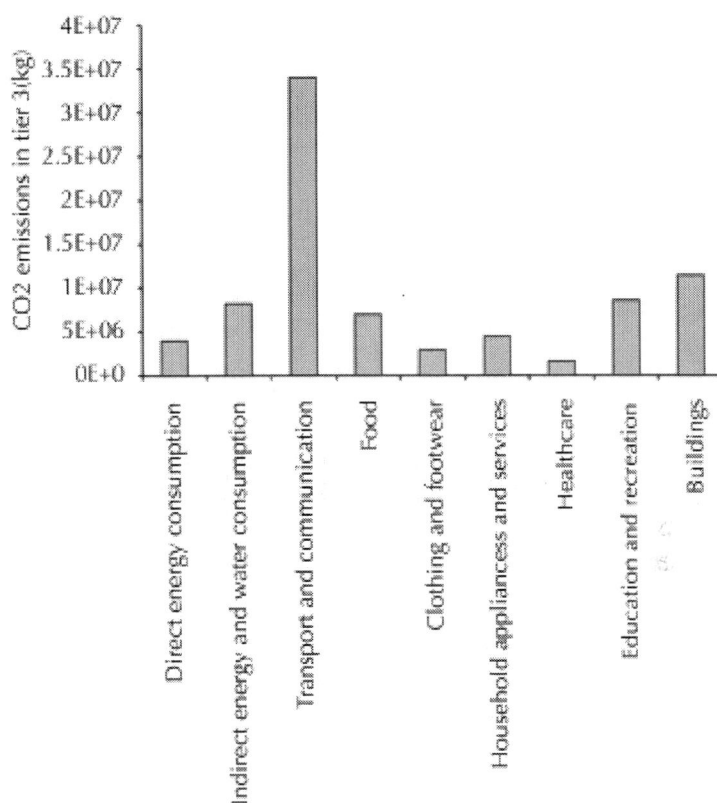

Figure 1: Total CO_2 emissions in Tier 3.

The buildings consume a large amount of materials, equipment, energy, and manpower at the stages of construction, fitment, outdoor facility construction, transportation, operation, waste treatment, property management, demolition, and disposal [64]. Due to a lack of data, only the main material consumption is considered in this study. Although this part occupies 14.11% of the total CO_2 emission, it is still smaller than the real value.

Energy consumption tends to increase along with income rise, which is confirmed by numerous studies [65, 66]. Thus, the main CO_2 emissions are from goods purchasing. The expenditure of health care is the smallest, which is mainly due to the age structure present in this community.

Comparison with Nanjing Community

There is a previous study on the typical community of Nanjing-Zhujiang Road Community (termed as Site A) [67]. Per capita CO_2 emissions of Site A from electricity, natural gas, and petrol consumptions are 1144.5 kg, 48.7 kg, and 540.1 kg, while in our case are 974.19 kg, 374.19 kg, and 893.55 kg, respectively. CO_2 emission from electricity of Beijing case is 14.88% lower than that of Site A. The younger residents in Beijing community have a better sense of energy conservation and usually prefer energy saving appliances. The CO_2 emission from natural gas of Beijing case is nearly seven times higher than that of Site A because space heating in northern China contributes the most while people do not have heating services in southern China like Nanjing. Meanwhile, the CO_2 emissions from petrol consumption of Beijing case are 65.44% higher than that of Site A due to longer distance between working place and home in Beijing compared to Nanjing. Particularly, our case considers the total emissions embodied in the supply chain, which is often significantly underestimated by the previous studies.

CONCLUSIONS

In this paper, a new carbon accounting framework, that is, three-tier accounting method, was established to estimate the total embodied CO_2 emissions of urban community. The carbon emissions and the mitigation potential were quantified according to the proposed accounting framework. From the results we can obtain that in the concerned community only 11.46% of Tier 3 are captured by Tier 2. The major carbon source is the total supply chain emissions embodied in the consumption of goods and activities. The results also indicated that for the total CO_2 emissions, the top 3 emission items are transport and communication (41.36%), buildings (14.11%), and education and recreation (10.41%).

As can be seen, the mitigation emphases should be placed on Tier 3. Two major suggestions are there by provided to realize the optimal resource allocation and further promotion for such communities. One is that we should strengthen the promotion of energy-efficient or green building and pay more attention to the renewable energy appliances such as solar energy water heater. The architectural of the houses should also be improved to reduce energy consumption of lightning and space heating. On the other hand, due to public transportation, the reconstruction of the urban public transportation is needed to reduce CO_2 emissions caused by the huge growth of private car ownership.

ACKNOWLEDGMENTS

This study was supported by the Key Program of National Natural Science Foundation (no. 50939001), National Science Foundation for Innovative Research Group (no. 51121003), National Natural Science Foundation (no. 41271543), and Program for New Century Excellent Talents in University (NCET-09-0226). The authors are especially grateful for the financial support from Beijing Development Area Co. Ltd.

REFERENCES

1. S. Q. Chen, B. Chen, and D. Song, "Life-cycle energy production and emissions mitigation by comprehensive biogas-digestate utilization," Bioresource Technology, vol. 114, pp. 357–364, 2012.

2. J. Wang, M. R. Su, B. Chen, S. Q. Chen, and C. Liang, "A comparative study of Beijing and three global cities: a perspective on urban livability," Frontiers of Earth Science in China, vol. 5, no. 3, pp. 323–329, 2011.

3. B. Chen, G. Q. Chen, and Z. F. Yang, "Exergy-based resource accounting for China," Ecological Modelling, vol. 196, no. 3-4, pp. 313–328, 2006.

4. B. Chen and G. Q. Chen, "Ecological footprint accounting based on emergy—a case study of the Chinese society," Ecological Modelling, vol. 198, no. 1-2, pp. 101–114, 2006.

5. B. Chen, G. Q. Chen, Z. F. Yang, and M. M. Jiang, "Ecological footprint accounting for energy and resource in China," Energy Policy, vol. 35, no. 3, pp. 1599–1609, 2007.

6. G. Q. Chen and B. Chen, "Resource analysis of the Chinese society 1980–2002 based on exergy—part 1: fossil fuels and

energy minerals," Energy Policy, vol. 35, no. 4, pp. 2038–2050, 2007.

7. B. Chen and G. Q. Chen, "Resource analysis of the Chinese society 1980 –2002 based on exergy—part 2: renewable energy sources and forest," Energy Policy, vol. 35, no. 4, pp. 2051–2064, 2007.

8. B. Chen and G. Q. Chen, "Resource analysis of the Chinese society 1980–2002 based on exergy—part 3: agricultural products," Energy Policy, vol. 35, no. 4, pp. 2065–2078, 2007.

9. B. Chen and G. Q. Chen, "Resource analysis of the Chinese society 1980–2002 based on exergy—part 4: fishery and rangeland," Energy Policy, vol. 35, no. 4, pp. 2079–2086, 2007.

10. B. Chen and G. Q. Chen, "Resource analysis of the Chinese society 1980–2002 based on energy—part 5: resource structure and intensity," Energy Policy, vol. 35, no. 4, pp. 2087–2095, 2007.

11. Z. M. Chen, B. Chen, J. B. Zhou et al., "A vertical subsurface-flow constructed wetland in Beijing,"Communications in Nonlinear Science and Numerical Simulation, vol. 13, no. 9, pp. 1986–1997, 2008.

12. X. H. Zhang, H. W. Zhang, B. Chen, G. Q. Chen, and X. H. Zhao, "Water resources planning based on complex system dynamics: a case study of Tianjin city," Communications in Nonlinear Science and Numerical Simulation, vol. 13, no. 10, pp. 2328–2336, 2008.

13. M. M. Jiang, J. B. Zhou, B. Chen, and G. Q. Chen, "Emergy-based ecological account for the Chinese economy in 2004," Communications in Nonlinear Science and Numerical Simulation, vol. 13, no. 10, pp. 2337–2356, 2008.

14. B. Chen, Z. M. Chen, Y. Zhou, J. B. Zhou, and G. Q. Chen, "Emergy as embodied energy based assessment for local sustainability of a constructed wetland in Beijing," Communications in Nonlinear Science and Numerical Simulation, vol. 14, no. 2, pp. 622–635, 2009.

15. B. Chen, G. Q. Chen, F. H. Hao, and Z. F. Yang, "The water resources assessment based on resource exergy for the mainstream Yellow River," Communications in Nonlinear Science and Numerical

Simulation, vol. 14, no. 1, pp. 331–344, 2009.

16. B. Chen and G. Q. Chen, "Emergy-based energy and material metabolism of the Yellow River basin,"Communications in Nonlinear Science and Numerical Simulation, vol. 14, no. 3, pp. 923–934, 2009

17. X. M. Wei, B. Chen, Y. H. Qu, C. Lin, and G. Q. Chen, "Emergy analysis for "Four in One" peach production system in Beijing," Communications in Nonlinear Science and Numerical Simulation, vol. 14, no. 3, pp. 946–958, 2009.

18. B. Chen, G. Q. Chen, F. H. Hao, and Z. F. Yang, "Exergy-based water resource allocation of the mainstream Yellow River," Communications in Nonlinear Science and Numerical Simulation, vol. 14, no. 4, pp. 1721–1728, 2009. View at Publisher · View at Google Scholar · View at Scopus

19. Q. Yang, B. Chen, X. Ji, Y. F. He, and G. Q. Chen, "Exergetic evaluation of corn-ethanol production in China," Communications in Nonlinear Science and Numerical Simulation, vol. 14, no. 5, pp. 2450–2461, 2009.

20. J. B. Zhou, M. M. Jiang, B. Chen, and G. Q. Chen, "Emergy evaluations for constructed wetland and conventional wastewater treatments," Communications in Nonlinear Science and Numerical Simulation, vol. 14, no. 4, pp. 1781–1789, 2009.

21. G. Y. Liu, Z. F. Yang, B. Chen et al., "Emergy-based urban ecosystem health assessment: a case study of Baotou, China," Communications in Nonlinear Science and Numerical Simulation, vol. 14, no. 3, pp. 972–981, 2009.

22. M. R. Su, Z. F. Yang, and B. Chen, "Set pair analysis for urban ecosystem health assessment,"Communications in Nonlinear Science and Numerical Simulation, vol. 14, no. 4, pp. 1773–1780, 2009.

23. Y. Zhang, Y. W. Zhao, Z. F. Yang, B. Chen, and G. Q. Chen, "Measurement and evaluation of the metabolic capacity of an urban ecosystem," Communications in Nonlinear Science and Numerical Simulation, vol. 14, no. 4, pp. 1758–1765, 2009.

24. L. X. Zhang, B. Chen, Z. F. Yang, G. Q. Chen, M. M. Jiang, and G. Y. Liu, "Comparison of typical mega cities in China using emergy synthesis," Communications in Nonlinear Science and

Numerical Simulation, vol. 14, no. 6, pp. 2827–2836, 2009.

25. M. M. Jiang, J. B. Zhou, B. Chen et al., "Ecological evaluation of Beijing economy based on emergy indices," Communications in Nonlinear Science and Numerical Simulation, vol. 14, no. 5, pp. 2482–2494, 2009.

26. Z. F. Yang, T. Sun, B. S. Cui, B. Chen, and G. Q. Chen, "Environmental flow requirements for integrated water resources allocation in the Yellow River Basin, China," Communications in Nonlinear Science and Numerical Simulation, vol. 14, no. 5, pp. 2469–2481, 2009.

27. X. Zhao, B. Chen, and Z. F. Yang, "National water footprint in an input-output framework—a case study of China 2002," Ecological Modelling, vol. 220, no. 2, pp. 245–253, 2009.

28. G. Q. Chen, M. M. Jiang, J. B. Zhou, B. Chen, Z. F. Yang, and X. Ji, "Exergetic assessment for ecological economic system: Chinese agriculture," Ecological Modelling, vol. 220, no. 3, pp. 397–410, 2009.

29. Z. F. Cai, Q. Yang, B. Zhang, H. Chen, B. Chen, and G. Q. Chen, "Water resources in unified accounting for natural resources," Communications in Nonlinear Science and Numerical Simulation, vol. 14, no. 9-10, pp. 3693–3704, 2009. X. Ji, G. Q. Chen, B. Chen, and M. M. Jiang, "Exergy-based assessment for waste gas emissions from Chinese transportation," Energy Policy, vol. 37, no. 6, pp. 2231–2240, 2009.

30. G. Q. Chen and B. Chen, "Extended-exergy analysis of the Chinese society," Energy, vol. 34, no. 9, pp. 1127–1144, 2009.

31. M. R. Su, Z. F. Yang, B. Chen, and S. Ulgiati, "Urban ecosystem health assessment based on emergy and set pair analysis—a comparative study of typical Chinese cities," Ecological Modelling, vol. 220, no. 18, pp. 2341–2348, 2009.

32. G. Y. Liu, Z. F. Yang, B. Chen, and S. Ulgiati, "Emergy-based urban health evaluation and development pattern analysis," Ecological Modelling, vol. 220, no. 18, pp. 2291–2301, 2009.

33. Z. F. Yang, M. M. Jiang, B. Chen, J. B. Zhou, G. Q. Chen, and S. C. Li, "Solar emergy evaluation for Chinese economy," Energy Policy, vol. 38, no. 2, pp. 875–886, 2010.

34. Z. M. Chen, G. Q. Chen, J. B. Zhou, M. M. Jiang, and B. Chen, "Ecological input-output modeling for embodied resources and emissions in Chinese economy 2005," Communications in Nonlinear Science and Numerical Simulation, vol. 15, no. 7, pp. 1942–1965, 2010.

35. IPCC, "Climate change 2007," Comprehensive Report, 2007.

36. S. Q. Chen and B. Chen, "Network environ perspective for urban metabolism and carbon emissions: a case study of Vienna, Austria," Environmental Science and Technology, vol. 46, no. 8, pp. 4498–4506, 2012.

37. Y. X. Dai, "The necessity and governance model of developing low carbon city in China," China Population, Resources and Environment, vol. 19, no. 3, pp. 12–17, 2009 (Chinese).

38. Z. P. Xin and Y. T. Zhang, "Low carbon community and its practice," Urban Issues, no. 10, pp. 91–95, 2008 (Chinese).

39. ARUP, "Beijing changxindian low carbon community concept plan," Urbanism and Architecture, no. 2, pp. 44–46, 2010 (Chinese).

40. L. Middlemiss, "Influencing individual sustainability: a review of the evidence on the role of community-based organisations," International Journal of Environment and Sustainable Development, vol. 7, no. 1, pp. 78–93, 2008.

41. E. Heiskanen, M. Johnson, S. Robinson, E. Vadovics, and M. Saastamoinen, "Low-carbon communities as a context for individual behavioural change," Energy Policy, vol. 38, no. 12, pp. 7586–7595, 2010.

42. Department of Energy and Climate Change (DECC), Low Carbon Transition Plan, Department of Energy and Climate Change (DECC), London, UK, 2009.

43. INC/FCCC, United Nations Framework Convention on Climate Change, 2000.

44. EPD and EMSD, Guidelines to Account for and Report on Greenhouse Gas Emissions and Removals for Buildings in Hong Kong, 2008.

45. CBEEX, Panda Standard, 2009.

46. J. Munksgaard, K. A. Pedersen, and M. Wien, "Impact of

household consumption on CO_2 emissions,"Energy Economics, vol. 22, no. 4, pp. 423–440, 2000.

47.   S. Pachauri and D. Spreng, "Direct and indirect energy requirements of households in India," Energy Policy, vol. 30, no. 6, pp. 511–523, 2002.

48.   Z. Lu, R. Matsuhashi, and Y. Yoshida, "Direct and indirect impacts of households by region of China on CO_2 emissions," Environmental Informatics Archives, vol. 5, pp. 214–223, 2007.

49.   H. He, "Research on green space carbon sink of residential area in South China and its application in carbon budget of residential area for whole life circle" (Chinese), 12, Chongqing University, 2010.

50.   H. S. Matthews, C. T. Hendrickson, and C. L. Weber, "The importance of carbon footprint estimation boundaries," Environmental Science and Technology, vol. 42, no. 16, pp. 5839–5842, 2008.

51.   Y. A. Huang, C. L. Weber, and H. S. Matthews, "Categorization of scope 3 emissions for streamlined enterprise carbon footprinting," Environmental Science and Technology, vol. 43, no. 22, pp. 8509–8515, 2009.

52.   H. N. Larsen and E. G. Hertwich, "The case for consumption-based accounting of greenhouse gas emissions to promote local climate action," Environmental Science and Policy, vol. 12, no. 7, pp. 791–798, 2009.

53.   L. P. Ju and B. Chen, "An input-output model to analyze sector linkages and CO_2 emissions," Procedia Environmental Sciences, vol. 2, pp. 1841–1845, 2010.

54.   A. Tukker, "Life cycle assessment as a tool in environmental impact assessment," Environmental Impact Assessment Review, vol. 20, no. 4, pp. 435–456, 2000.

55.   EN ISO 14040, "Environmental management—life cycle assessment—principles and framework," 2006.

56.   Y. Chang, R. J. Ries, and Y. W. Wang, "The embodied energy and environmental emissions of construction projects in China: an economic input-output LCA model," Energy Policy, vol. 38, no. 11, pp. 6597–6603, 2010.

57. W. W. Leontief, "Quantitative input and output relations in the economic systems of the United States," Review of Economics and Statistics, vol. 18, no. 3, pp. 105–125, 1936.

58. J. Munksgaard and K. A. Pedersen, "CO_2 accounts for open economies: producer or consumer responsibility?" Energy Policy, vol. 29, no. 4, pp. 327–334, 2001.

59. IPCC National Greenhouse Gas Inventories Programme, 2006 IPCC Guidelines For National Greenhouse Gas Inventories, vol. 2 of Energy, Institute for Global Environmental Strategies, Hayama, Japan, 2006.

60. Chinese Academy for Environment Planning, Analysis and Forecast of Environment and Economic about the Key Industries of Energy Saving and Emission Reduction in China From 2009–2020, China Environmental Science Press, Beijing, China, 2009.

61. G. Q. Chen, "Systems cart on metrics for building" (Chinese), Beijing, China, 2010.

62. C. L. Weber and H. S. Matthews, "Quantifying the global and distributional aspects of American household carbon footprint," Ecological Economics, vol. 66, no. 2-3, pp. 379–391, 2008.

63. G. Q. Chen, H. Chen, Z. M. Chen et al., "Low-carbon building assessment and multi-scale input-output analysis," Communications in Nonlinear Science and Numerical Simulation, vol. 16, no. 1, pp. 583–595, 2011.

64. L. D. Shorrock, "Identifying the individual components of United Kingdom domestic sector carbon emission changes between 1990 and 2000," Energy Policy, vol. 28, no. 3, pp. 193–200, 2000

65. L. J. Schipper, R. Haas, and C. Sheinbaum, "Recent trends in residential energy use in OECD countries and their impact on carbon dioxide emissions: A comparative analysis of the period 1973–1992,"Mitigation and Adaptation Strategies for Global Change, vol. 1, no. 2, pp. 167–196, 1996.

66. Z. H. Gu, Q. Sun, and R. Wennersten, "Impact of urban residences on energy consumption and carbon emissions–an investigation in Nanjing," China, 2012.

Biodiesel from Oilseeds in the Canadian Prairies and Supply-Chain Models for Exploring Production Cost Scenarios: A Review

Nathaniel K. Newlands[1]
and Lawrence Townley-Smith[2]

[1]Environmental Health, Agriculture and Agri-Food Canada, Lethbridge Research Centre, Lethbridge, AB, Canada T1J 4B1

[2]Agri-Environmental Service Branch (AESB), Agriculture and Agri-Food Canada, 1800 Hamilton Street, Regina, SK, Canada S4P 4L2

ABSTRACT

Canada recently implemented a federal mandate of 2% of renewable content in diesel fuel and heating oil. Federal-level biofuel strategy is currently more geared to bioethanol, as nonfood oils continue to be more cost-competitive and canola seeded area is forecast to increase 10% as a new record due to strong prices and high expected yields. Increasing focus is therefore being placed on alternative oilseeds as nonfood crops for biodiesel and their ability to adapt to the semiarid conditions of the Canadian Prairies and provide benefits in nutrient and water-use efficiency when introduced into the crop rotation. Systems engineering and supply-chain modeling and optimization will have an increasingly important role in decision making for designating supply units, the linkage of processes and chains, and biorefinery system design. However, current models require further enhancement to address current challenging questions: (1) changing spatial considerations (e.g., land use and suitability for feedstocks), (2) changing temporal dynamics of supply and risk of climate extreme impacts on transportation networks (road, rail, pipeline), price volatility, changes in policy targets and subsidy regimes, process technological change, and multigenerational biorefinery systems engineering advancements. Greater integration internationally in model development and testing would improve sensitivity and reliability in their system-level predictions and forecasts.

INTRODUCTION

Biofuel is fuel that is derived from biomass or living organisms and/or their metabolic byproducts. It is termed renewable (unlike fossil fuel (petroleum, coal) and nuclear energy sources) because it produces electrical, thermal, and/or mechanical energy at rates that are faster than the rate at which its resource base is consumed. Canada's annual primary energy supply is roughly 11 exajoules of which roughly 17% is renewable (i.e., 11% from hydroelectricity and 6% from biomass) [1]. Nonetheless, renewable bioenergy supplied from agricultural reigons and forest wastes (with contributions from industrial, municipal solid waste, and sewage biogas), energy crops, wind, and solar sources continues to increase. Currently, the pulp and paper and forest-product industries recycle half of their total energy use by converting bioenergy into electricity, steam, and heat, while fuelwood and gas from landfills are used in heating residential spaces. Bioenergy is currently transformed into biofuel for generating power as urban and rural electricity, heating water and spaces, and transportation.

Biofuel production, to be a viable renewable source of energy supply, must provide a net energy gain, have environmental benefits, be economically competitive, and be producible in large quantities, without reducing food supplies [2]. The use of biofuel in Canada's transportation is of special interest given that this sector contributes about 28% of national carbon dioxide (CO_2) emissions, in addition to reducing environmental air quality [3]. Here, displacing fossil fuel with biofuel would reduce net emissions of carbon into the atmosphere and would help to mitigate environmental impacts of increased atmospheric greenhouse gases (GHGs). The agricultural sector also has a direct role in some of the newly implemented biobased energy systems because farm products are the primary input into many of these systems, including grains (ethanol), oilseeds (biodiesel), waste products (biogas), cellulosic materials (ethanol), and woody biomass (heat energy, biogas). In 2011, Canada implemented a federal mandate of 2% of renewable content in diesel fuel and heating oil that sets a target of 600 ML yr^{-1} of domestic production from current 2011 levels forecasted at 158 ML yr^{-1} (that have attained a 13% increase from 2010 levels of 140 ML yr^{-1} or 0.1 GL yr^{-1}). This compares to current biodiesel production of 6.1 GL yr^{-1} in the European Union and 1.9 GL yr^{-1} (505 mgy) in the United States. In the U.S. there are 191 biodiesel operating facilities having a capacity of 10.6 GL yr^{-1} (i.e., 2.8 billion gallons yr^{-1} (bgy), US EPA, 2010). A biorefinery in Saskatchewan (Canada Bioenergy/ADM plant situated in Lloydminster) is currently under construction and is anticipated to raise Canada's biodiesel production capacity by 225 ML to a total of 475 ML in 2012 mainly from 10 biorefineries in the Canadian Prairies (western Canada) and another 4 in Ontario and Quebec (eastern Canada).

Under a multicriteria objective of minimizing production cost, reducing net GHG emissions, and to minimize soil degradation and other harmful environmental impacts, a systems perspective is required to link economic with environmental considerations—involving detailed analysis and modeling (i.e., life-cycle assessment) across biofuel supply chains. Such assessment-optimization-simulation quantitative methodologies can then more reliably assess if a biofuel provides real attainable benefits when displacing fossil fuels. Current systems-level research on biofuels (biodiesel and bioethanol) is directed to better assessing: the potential impacts of weather on feedstock supply and collection, more reliable forecasting of in-season feedstock yields, evaluating scenarios that explore climate trends and extremes on feedstock redistribution in relation to differential landscape suitability, assessing net-energy savings provided by coproducts and their industrial and commercial use within a range of potential industry and consumer markets, and assessing potential benefits of improved processing technology in reducing net GHG's. Amongst all these considerations, feedstock availability and cost remain the driving factors (i.e., 35–50% of production costs) having a major influence in determining where biorefineries are located as well as the rate at which the bioindustry expands. Existing models for biofuel supply-chain analysis, optimization, and simulation of future scenarios each have their various strengths and weaknesses in terms of flexibility/ease in enabling further model refinements and reliability testing. These models aim to explore process dynamics and to provide key insights to challenging questions, such as the following. (1) What are the key cost barriers in biofuel production and expansion? (2) What supply system/infrastructure improvements are needed and how to cope with such changes? (3) What are the technological advancements that will help to achieve government policy targets? (4) How to simultaneously maintain supply for multiple biorefineries, that is, systems of biorefineries and multicommodity supply-chains? Here, we focus attention on a selected set of 6 models representative of "state-of-the-art" biofuel supply chain analysis and provide key recommendations for improving supply-chain modeling both from a research (i.e., modeling) and their real-world multigenerational, multiproduct biorefinery supply perspective.

MATERIALS AND METHODS

Data Sources

Available data compiled on oilseed and biodiesel production was obtained from market analysis reports and forecasts produced by Agriculture and Agri-Food Canada's Crop Market Information reporting [11] and the Global Agricultural Information Network (GAIN), United States Department of Agriculture (USDA) Foreign Agricultural Service. Information from technical reports and the published scientific literature was compiled for detailing the structure, attributes, and characteristics of the supply chain models (refer to references linked within Table 2).

Oilseed crop yield (kg/ha, bushels/ha) and cropped area (ha) within Census of Agriculture Regions (CARs) (Statistics Canada, Field

Crop Reporting Series) were used to generate the spatial distribution of oilseed production as the product of yield and area within each CAR polygon. This data is from all farms enumerated through the census and is weighted in order to produce unbiased level indicators which are representative of the population. These level indicators then undergo a validation process, based on subject matter analysis and consultation with provincial statisticians, before a final estimate is published. Coefficients of variation (CV) for specialty crops and small areas of major crops lie within 5% to 15%.

Historical, interpolated climate data at the 10 km, daily resolution has been validated across Canada and released to the public as a national agri-geomatics database by Agriculture and AgriFood Canada [12]. This high-resolution database was used to generate the frost kills spatial maps, assuming first frost kill occurs when air temperature is <−2°C.

Overview of Supply-Chain Structure and Key Variables

The biodiesel supply chain consists of 5 major segments: feedstock production, logistics, conversion, distribution, and end use (Figure 1). Feedstock production considers land use, land suitability, and crop area expansion and intensification. In Canada, lower tillage intensity and absence of liming reduce the energy required to grow oilseed crops; tillage systems applied are conventional till, minimum, and no-till. The next segment involves logistical considerations linked with supply scheduling, namely, collection, pretreatment/drying, storage, and transportation. Here, for example, greenhouse gas (GHG) policies of carbon taxation, cap and trade, and carbon subsidies have the potential to significantly change not only scheduling of multiple feedstocks as part of segment 2 in the supply chain but also the selection of feedstocks (segment 1) and the design of biodiesel production facilities (segments 3, 4, and 5) [13]. Fertilizer and herbicide treatments for oilseeds are relatively uniform, specific to a given oilseed, due to the crop sequencing effect. This effect is the benefit of growing specific crops in the same field year after year arising from the maintenance of soil physical, chemical, and biological properties coupled with agronomic practices that promote nutrient and water-use efficiency and the control of diseases and pests.

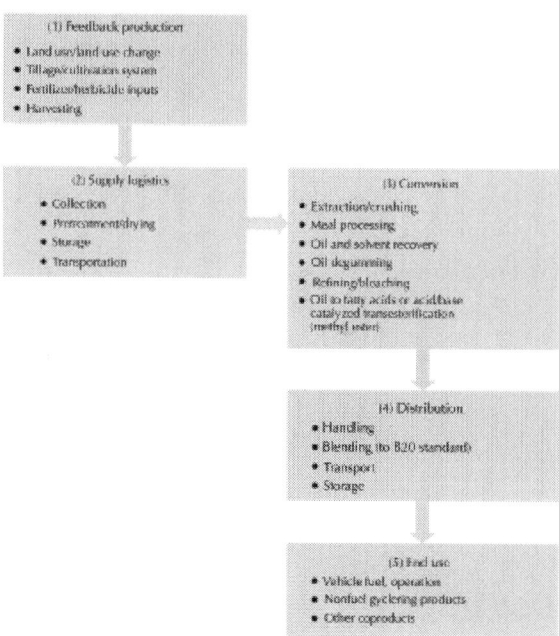

Figure 1: Generalization of the biodiesel supply chain with five major segments: feedstock production, logistics, conversion, distribution, and end use.

A recent review of energy crops in rotation reveals that conventional crops can benefit from the introduction of oilseed crops for biodiesel production in the rotation [14]. Introduction of oilseeds into the rotation would impact segment 1 and may also include adjustments in harvesting methods. This situation is commonly considered in the case of bioethanol where reduced row space and higher cellulosic material content require harvesting technological adaptations, but less so for the case when oilseeds are introduced into the rotation in new areas and for land with varying topology. Diverse benefits of alternative oilseeds introduced into rotation, which are still able to produce high yields with high oil content, likely will relate to advantages in use of water and nutrients across the soil profile—especially in semiarid regions, such as the Canadian Prairies. With flax having a typically shallower rooting system than other oilseeds (e.g., safflower has a much deeper rooting), introducing other oilseeds in rotation could take advantage of accumulated soil nutrient and changes in soil-water content through the growing season.

Pre-treatment involved in segment 2 (supply logistics) considers preparation of biomaterial in relation to transport capacity and desired moisture content and densification level as well as performance limits for reactors and separators linked with conversion processes within segment 3. Recent life-cycle assessment (LCA) for biodiesel energy balance shows an increase of 14% in energy gain for no-till compared to conventional tillage [15]. Energy inputs required to grow and convert canola to biodiesel are higher than for soybean, reflecting higher crop production inputs (i.e., fertilizer) and larger amounts of oil required for transesterification (conversion segment 3). Second-generation (2G) biorefinery design considerations are complex and involve multiple biochemical conversion processes—for example, 2G biodiesel combines gasification with syngas and Fischer-Tropsch synthesis to produce liquefied petroleum gas (LPG), naphtha, jet oil, and lubricants. Naphtha is important for the mining and chemical industry as it is used in diluents and solvents, respectively. It helps to meet viscosity and density requirements for pipeline transportation of liquid fuel material. Added hydrogenation to "green diesel" produces fuel blends with enhanced energy and GHG benefits compared with conventional biodiesel and to help meet end-user demands (segment 5) [16].

Optimization of Multi-criteria Objectives

The complexities involved in supply-chain modeling are optimized to meet various target objectives. Typically, total production cost over the chain is chosen as the main objective or utility function from which constraints are based. This is in part due to the greater availability of supporting data for setting look-up tables (LUT) and other input data required for modeling against this objective. However, supply-chain models are expanding their efforts to explore scenarios around near-feasible solutions against net-CO_2 equivalent emission (i.e., net GHG's) objectives. Supply-chain models typically utilize idealized, grid-based coarse-grained spatial distribution data on feedstock distribution which offers considerable advantages for solving models using available linear and nonlinear programming (LP) optimization algorithms and exploring feasible and scenario-solution sets.

RESULTS

Regional Perspective: Oilseed Production and the Canadian Prairies

Globally, soybean (Glycine L.) accounts for 60% of oilseed production, with cottonseed (Gossypium herbaceum L.) providing 10%, closely behind napus canola (Brassica napus L.) and rapa canola (Brassica rapa L.) "rapeseed." Given the regional climate and agronomic factors related to favorable oilseed production, oilseed production in Canada is dominated by canola and flax (Linum usitatissimum L.) in the Canadian Prairies and by soybean in eastern Ontario and Quebec. The spatial distributions of major oilseed crop production (tonnes) for 2008 is shown in Figure 2, illustrating the overriding importance of land suitability, favourable climate, and other agronomic considerations required to grow oilseeds under Canadian conditions. Area polygons are based on Census of Agriculture Regions (CAR's) crop reporting subdivisions for the Census of Agriculture conducted by Statistics Canada. The year 2008 was selected here to illustrate actual production for canola and flax during what was a particularly extreme climatic year—the Canadian Prairies experienced one of the worst droughts in decades, ravaging grain and oilseed crops, including hay and pasture. Northwestern Alberta experienced rainfall totals were 60 mm or 40% less than normal—its driest period in more than 40 years, with dryness becoming more extreme further southward. Daytime instability due to weaker weather frontal systems moving across the Prairies bringing warm, moist air from the Pacific reacted violently with the predominantly cooler air than normal residing over the Prairie region—producing intense storm and tornado activity. The summer of 2008 was the driest in 125 years in this growing region. The drought tolerance of both canola and flax is evident from attained production levels during such extreme conditions.

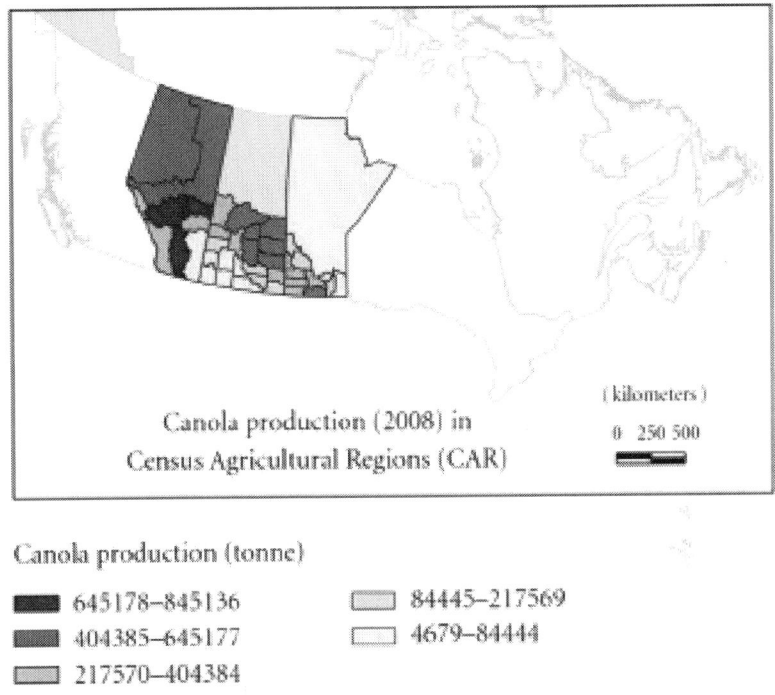

Canola production (tonne)

- ■ 645178–845136
- ■ 404385–645177
- □ 217570–404384
- □ 84445–217569
- □ 4679–84444

Figure 2: Spatial distribution of production of major oilseed crops (tonnes) (i.e., canola and flaxseed in Western Canada and soybean in Eastern Canada) for 2008. Area polygons are based on Census of Agricultural Regions (CARs) crop reporting subdivisions of Statistics Canada.

Canola (short for "Canadian oil") was developed by plant breeders in the Canadian Prairies (Saskatchewan/Manitoba) during the 1960–70s through conventional plant-breeding experiments [17]. Previously, the main impediment to growing canola/rapeseed varieties for food/edible oil production was due to its significant level of erucic and eicosenoic acids (nutritionally undesirable) and presence of sulphur compounds (glucosinolates) which impart sharp flavour characteristics (not favorable for wide human and livestock consumption). While this impediment was overcome by Canada and has made canola a major oilseed globally, there are other impediments facing the use of other oilseeds and the tradeoff in their use for food versus fuel. For example, one particular advantage of cottonseed oil that is increasingly being recognized is that it is more saturated than soybean oil and can be used to create oil blends or reduce the need for hydrogenation that creates transfatty acids. However, there are also increasing concerns in its use in the food industry given increasing evidence of its proteins causing allergic sensitivity and more serious reactions. The addition of cottonseed as the oil of choice (as when peanut was used) in manufacturing penicillin and vaccines may have contributed to major outbreaks of allergies and anaphylaxis in North America—and one reason why cottonseed although a more inexpensive oil, may be less favorable for food consumption as a cooking oil (i.e., than canola which has recognized health benefits as a source of unsaturated fat/vegetable oil) and better for biofuel or other bioindustrial production (e.g., detergents, soaps, fertilizers, inks, lubricants). Clearly, beneficial societal tradeoffs in the use of oilseeds for food or fuel production need to be better understood—and here the use of alternative nonfood oilseeds may have distinct advantages that Canadian oilseed producers and production biorefineries could exploit and expand upon.

Alternative oilseed crops have also been evaluated under the semiarid, short-growing season of the Canadian Prairies, such as juncea canola (Brassica juncea L.), Ethiopian mustard (Brassica carinata L.), oriental mustard (Brassica juncea L.), yellow mustard (Sinapis alba L.), and camelina (Camelina sativa L.) [18]. In particular, Camelina is considered more drought and heat tolerant and less susceptible to dominant pests. Camelina needs little water or nitrogen to flourish, and can be grown on marginal agricultural land and does not compete with food crops—so can be introduced as a rotation crop for wheat, increasing the health of the soil. Recent experimental studies in the Canadian Prairies on oilseeds indicate that crop emergence and growth were generally good for all oilseed crops, but soybean did not fully mature at some locations. Considering yield and oil concentration, the alternative oilseed crops exhibiting the most potential for biodiesel feedstock were camelina, flax, rapa canola and oriental mustard, respectively. Oils of all crops were easily converted to biodiesel and quality analyses indicated that all crops would be suitable for biodiesel feedstock with the addition of antioxidants that are routinely utilized in biodiesel fuels. Reported findings from a multisite field study support that Ethiopian mustard has a comparable yet a longer duration to maturity (102–115 days) than other competing major biodiesel feedstock sources such as camelina and canola varieties (87–95 days). This indicates additional risk in growing alternative oilseeds such as Ethiopian mustard, if the impact of late growing-season climate variability on emergence, growth, and yield potential is not sufficiently understood and accounted for. The long-term viability of using dedicated industrial oilseed crops such as Ethiopian mustard (Brassica carinata L.) as a renewable source of jet fuel will, therefore, require a detailed understanding of key meteorological factors influencing the adaptation of this crop to the Canadian Prairies [19]. Companies like Agrisoma Inc. are continuing to develop elite carinata lines that increase the vigor of this crop, increase its drought and heat tolerance (i.e., adaptability), and increase its resistance to Blackleg. Second-generation lines currently being developed aim to achieve a 10–20% increase in yield and significant increases in seed oil content, including enhancing its oil quality and crop productivity traits.

Current and Forecasted Biodiesel Production

For 2011-2012, seeded area for canola is forecast to increase 10% as a new record due to strong prices and high-expected yields (see Table 1). In contrast, flaxseed oil-seeded area is forecast to rise by 70%, but domestic use to fall by 36% due to high prices, in part due to a forecasted decline in area in summer fallow that considers that areas too wet to be seeded in 2010 will be seeded in 2011 [11]. However, this also assumes normal precipitation and crop quality, but yield forecasts are likely also particularly sensitive to spring moisture conditions. Moving from the feedstock to the biodiesel production perspective, most of the current forecasted increase in biodiesel production is based on potential production from rendered animal fats at 250 ML. Higher prices for oilseeds likely will hinder Canada's ability to supply the necessary supply feedstock to meet biodiesel targets. The Federal Government's biofuel strategy program is more geared to bioethanol reducing the ability to address limiting factors for biodiesel market growth—especially where sensitive trade-offs via subsidies between oil from crushing plants directed for biodiesel production versus human consumption are involved.

Table 1: Harvested area, yield, and total production of the two main oilseeds grown in the Canadian Prairies (canola and flaxseed) for 2009/10 and forecasts for 2010/11 and 2011/12 cropping years (i.e., August-July) (reproduced from available 2011 oilseed and biodiesel production market analysis reports and forecasts[†]) (k denotes 10^3, t metric tonnes, and M 10^6)

Oilseed crop	Variable	2009/10 actual	2010/11 forecast	2011/12 forecast
Canola	Area seeded (kt)	6,556	6,806	7,500
	Area harvested (kt)	6,105	6,514	7,368
	Yield (t/ha)	2.03	1.82	1.75
	Production (kt)	12.4	11.9	12.9
	Imports (kt)	128	250	125
	Total supply (kt)	14,206	14,239	14,125
	Average price ($/t)	426	540–580	500–540
Flaxseed	Area seeded (kt)	692	374	635
	Area harvested (kt)	623	353	630
	Yield (t/ha)	1.49	1.20	1.35
	Production (t)	930	423	850
	Imports	6	5	5
	Total supply	1,165	717	905
	Average price ($/t)	424	520–550	475–525
Biodiesel	Feedstock use (kt)	160	175	475
	Production (ML)	140	158	475
	Imports (ML)	56	57	145
	Exports (ML)	70	70	70
	Number of facilities (conventional blend capacity) (ML)	9 (186)	11 (207)	14 (558)

[†]http://www.agr.gc.ca/ under "Crops Market Information" and Global Agricultural Information Network (GAIN). Annual report for biofuels in Canada, 2011 US Department of Agriculture (USDA) Foreign Agricultural Service.

Table 2: Description and characteristics of four selected "state-of-the-art" models for optimizing biofuel production chains (i.e., biodiesel and/or ethanol) applicable to current decentralized configuration of multiple-biorefinery systems aimed at minimizing processing and production costs and enabling an enhanced level of strategic planning of future supply chain risks

Model source/ country	Model formulation/benefits	Applied constraints/ drawbacks
Parker et al. [4] (2010, United States)	(i) Mixed-integer LP[†] (ii) Maximizes annual revenue (iii) Considers feedstock handling efficiency/ loss, conversion cost and efficiency, and transportation costs (iv) Simulates industry-wide fuel production at fixed price, for generating regional and/or state level exploring supply-cost curve under different feedstock mixes (v) Model links with explicit feedstock spatial distributions	(i) Assumes value of all fuels have equal energy content (ii) Other than transport, does not consider many aspects of supply logistics (i.e., pretreatment, collection, storage) (iii) Impact of associated supply risks due to weather/ climate-related variability and extremes not taken into account (iv) Intraannual (i.e., within-year) supply dynamics not considered (v) Net greenhouse gas benefits of biofuel production from feedstock mixes not considered, yet potentially has large effect on future crop production feedstock/resource base

Dunnett et al. [5] (2008, United Kingdom)	(i) Single-commodity, discrete (i.e., grid-based) noninteger LP (ii) Minimizes annual production costs and system logistics (i.e., supply distance/time specific to each feedstock type) (iii) Considers rural, semirural, and urban region types, with assumed "industry of scale" cost reduction function (iv) Considers pretreatment efficiency, transfer speed, and loading/unloading time for each feedstock supply logistics (v) Flexible framework for including range of processing tasks, logistical modes, coproducts, and regional policy constraints as dynamic extensions for real-world case study	(i) Assumes 10% fractional availability of cropland as resource base (ii) No link to explicit feedstock spatial distribution considers idealized crop spatial distributions as regional typologies: centralized and corner point (iii) Lignocellulosic ethanol processing only currently simulated

"BioTrans" Van Tilberg et al. [6] (2005, European Union)	(i) Multicommodity, multistage, mixed-integer LP (ii) Annual time step, minimising production costs and system logistics (iii) Considers macroeconomic and technological projections in finding minimal cost allocations for supply chains (iv) Detailed consideration of conversion processes	(i) Operates on a country aggregation level. Input and projections can be set at national level and costs and production quantities determined (ii) Requires each country to have a complete production and supply chain with one production or processing facility of each type, so difficult to apply at regional level (iii) Impact of associated supply risks due to weather/climate-related variability and extremes not taken into account (iv) Intraannual (i.e., within-year/in-season) supply dynamics not considered
Huang et al. [7] (2010, United States)	(i) Multistage, mixed integer LP (ii) Annual time step, minimizes annual production costs and system logistics (iii) Considers transboundary of feedstock supply and associated outsourcing penalty costs (iv) Considers explicit feedstock distributions (v) Considers fixed candidate refinery locations (vi) Allows for increases in biorefinery capacity over time	(i) Developed for first application to lingocellulosic ethanol/biomass resources (ii) Landscape suitability/ratings specific to individual crops/feedstocks not currently considered, so only considers fixed set of candidate biorefinery locations (iii) Impact of associated supply risks due to weather/climate-related variability and extremes not taken into account

"IBSAL" Sokhansanj et al. [8] (2008, United States)	(i) Grid-based simulation (ii) Integrated biomass supply analysis and logistics model (IBSAL) (iii) Simulates the flow of biomass through collection, transport, storage, and preprocessing considers costs, energy, and net CO_2 emissions multiobjectives considers weather impacts on supply-chain logistics	(i) Relies on calibration of empirical relationships and other detailed look-up table (LUT) logistics operation data, that is, primarily focused on the front end of the biofuels supply chain at the local level (ii) No link to explicit feedstock spatial distribution (iii) Enables sensitivity analysis of input data in relation to empirically derived logistical functions, but does not optimize, so lacks ability to explore more full supply-chain and regional-scale scenarios
Newlands et al.[9, 10] (2010, Canada)	(i) Multicommodity, multistage, nonlinear LP (ii) Implemented with a global parameter optimization scheme for enhanced robustness as model complexity increases (e.g., multiple chains, national-scale application) (iii) Monthly time step, minimizes production costs and system logistics (iv) Considers multiple feedstocks/mix, multiple collection and single to multibiorefinery systems (v) Model links with explicit feedstock distributions (vi) Considers weather/ climate impacts as part of supply cost scenarios	(i) Developed for first application to lingocellulosic ethanol, subsequently being further applied/tested for forestry/ agricultural cross-sector biomass and biodiesel supply chains (ii) Landscape suitability/ ratings specific to individual crops/feedstocks not currently considered, so only considers fixed set of candidate biorefinery locations (iii) Does not currently include multicriteria as part of its optimization

† LP linear programming/optimization model.

The mix of feedstock that can most reliably supply the 2% biodiesel federal mandate is a major outstanding question. Rendered oils (yellow grease), animal fats (tallow), and palm oil (imported into Canada) are much cheaper oils that are priced at feed and industrial-use levels that strongly compete with use of canola for biodiesel. Also, canola (produced mainly in the Prairies, western Canada) and soybean (produced in Ontario, eastern Canada) are both high-priced food oils subject to international market prices and pressures. As their demand in food and biodiesel continues to rise (based on current forecasts), their price is likely to also increase. In contrast to canola soon to contribute over 50% of Canada's biodiesel, roughly 54% of biodiesel production in the United States is currently produced from soybean oil. Canada's and the United States' dominant use of canola, and soybean is related to the advantages of using higher-priced oilseeds in terms of quantity and quality of free glycerol or glycerin coproduct (roughly 10% of final product) as well as anticipated technological improvements in coproduct production processes.

High Spatial Variability and Volatility in Supply Risks

Future land-use simulations for the Canadian Prairies indicate an increase in the intensiveness of agricultural production of spring grains and oilseeds, and increased conversion of pasture acreage [20]. Such land conversion is predicted to occur heterogeneously, with CARs in Southern Saskatchewan losing less than 5% of their pasture acres, to North-central Alberta, where CARs are predicted to lose 20%. Not surprisingly, largest losses in pasture (i.e., the highest quality land available that is being used for growing hay and livestock pasture) and highest rates of land conversion are predicted for areas with low current land-use shares for intensive crops (i.e., wheat, canola, and flax). As more land converts for growing canola, for example, early frosts and wetter conditions can prove unfavorable at harvest as it can damage the seed so that it is unsuitable for human consumption— such that roughly 2–10% in any year may be currently available for biodiesel, due to such unfavorable conditions). Figure 3 shows the spatial distribution of early season (i.e., first) frost kill temperatures ($T < -2 °C$) across the Canadian Prairies at a spatial resolution of 10 km, for selected years. Frost kill is an indicator of higher probability in crop frost damage. These historical frost patterns show large interannual variability in frost kill and high unpredictability in predicting crop damage in any one year. As more land area is used to grow canola, given its high oil content, its storage needs to be kept dry and cool, which presents additional logistical challenges as oilseed crop area expansion and area intensification increase, alongside the demand for canola for both food and biodiesel use. Strategic policy biofuel initiatives that incorporate considerations of differential land suitability and its spatial heterogeneity in relation to conversion risks will likely be the ones that provide the greatest enhancement of supply-chain reliability, by reducing risk volatility, and increase the effectiveness of use of land for growing oilseeds used in both food and biodiesel supply-chain production. This is because allocating too much land for only one crop, such as canola, introduces known risks associated with monoculture—such as an increased potential risk of disease when occupying the full rotation and reduced soil fertility. High levels of climate and crop production spatial variability adds considerable complexity to modeling biodiesel supply-chain production—and requires an enhanced ability to forecast crop yield in-season. Such forecasting methodologies must consider potential impacts and risks of regional climate changes and extreme events for mapping changing land suitability (i.e., already spatially heterogeneous) and for forecasting different levels of supply risk volatility associated with different feedstock mixtures and their supply radii for regional biorefinery/ systems. Increasingly, the use of remote-sensing information to provide rapid, broad-based spatial coverages for identifying crop distribution and status through phenological development stages (i.e., in-season) is being integrated within agricultural crop forecasting system operational frameworks.

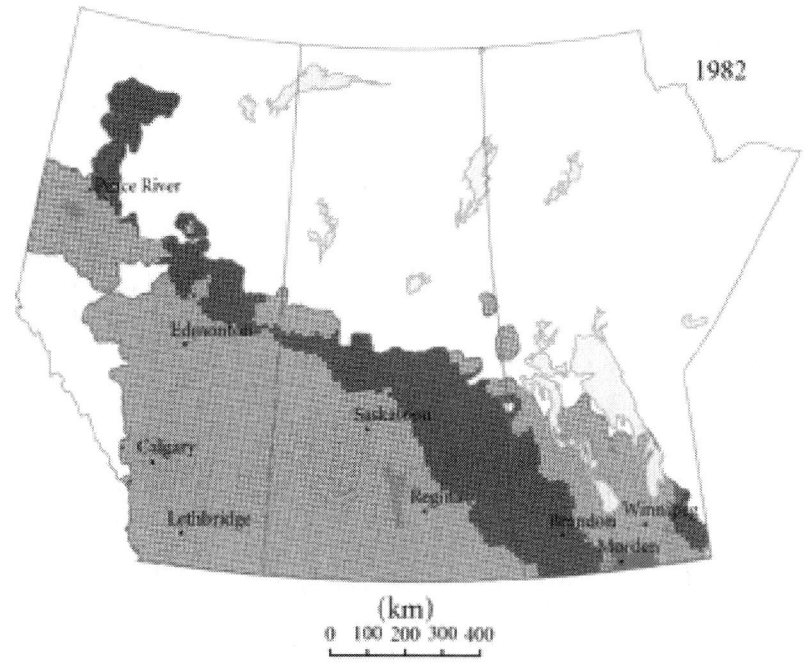

Figure 3: Spatial distribution of early-season (i.e., first) frost kill temperatures ($T < -2\,$°C) across the Canadian Prairies at a spatial resolution of 10 km, for selected years (frost kill is an indicator of higher probability in crop frost damage). These historical frost patterns show large interannual variability in frost kill and high unpredictability in predicting crop damage in any one year.

Supply-Chain Models: Benefits and Drawbacks

Six state-of-the-art models for optimizing supply chains (e.g., for biodiesel production) are profiled in Table 2. These models are currently being employed for exploring decentralized configurations of multiple-biorefinery systems aimed at minimizing processing and production costs [21–24]. While the level of detail in their input data, objective criteria, and spatial and temporal assumptions of supply chain dynamics differ, their structures all encompass the major segments of the supply-chain (as per Figure 1). Also, they collectively target the common modeling goal of enabling an enhanced level of strategic planning of future supply-chain risks. Assumed level of detail between the models, in part, reflects differing extents of data availability, whether they maintain linear assumptions on variables or not, accuracy of the core optimization numerical algorithm employed, application spatial scale (regional, national, transboundary), and desired level of flexibility for scenario-based simulation and the degree that they are designed for use by a range of stakeholders. For example, BioTrans (EU) [6] is particularly well designed for trans-boundary application, and IBSAL (US) [22] is well designed for supply logistics, and both have considerable flexibility and ease of use by stakeholders. Other models [4, 5, 7–10] emphasize supply network configuration design and are primarily research driven for exploring science-based or policy-setting scenarios. Overall, a summary of their characteristics reveals a common thread, whereby each model tends to focus on certain stages of the supply chain or central aspect to explore at greater detail building in data or concepts from more empirical sources—this identifies that there is a high potential for beneficial integration of improved segment design between the models. Moreover, substantial gains are likely if modelers were to increase the level of data and contextual-based information sharing. There are clear differences between the models in terms of their use of spatial information of feedstock supply (i.e., idealized, grid based, resolution scale). The evolution in the development of these models is clear: from considering single to mixes of different feedstocks, from cost only to multiobjectives, and from changing level of detail to best address changing scale of application (regional, national, trans-boundary). Nonetheless, there are several key areas where all the models require extension to better address data and model-based variability; namely, (1) changing spatial considerations (e.g., land-use and suitability for different feedstocks) (2) changing temporal dynamics of supply and risk due to climate-related impacts on transportation networks (road, rail, pipeline), market international pricing volatility, changes in policy targets and subsidy regimes as well as technological change associated with multi-generational biorefinery systems engineering advancements.

DISCUSSION

Without a sufficient level of spatial detail being considered, it may be difficult for supply-chain models to capture important spatial trends related to the impact of pests, climate extremes on crop productivity, or to obtain reliable estimates of feedstock supply and various mixes required by biorefinery systems. Current supply-chain models need to better incorporate geospatial information to enable more resolved ability to explore the effects of land use change. Important spatial effects that need to be considered include the increasing encroachment of crops on more marginal lands, as well as, crop intensification dependent on land suitability, climate fluctuations and its regionally-varying impact. With spatial enhancement, these models can be used to provide key statistical and scenario-based insights on system adaptation in response to favourable meteorological conditions for growing new oilseed alternatives for emerging markets such as jet fuel, lubricants, plastics, and other bio-based materials. How alternative oilseeds introduced into the rotation can make more use of available soil nutrients and retained water in relation to their rooting depth and how further intensification could lead to unfavourable consequences currently provided by crop sequencing are important questions that would assist in designing more realistic supply-chain scenarios for biodiesel in the Canadian Prairies. A review of a representative set of state-of-the-art supply-chain models reveals several key areas where all the models require extension to better address data and model-based variability; namely, (1) changing spatial considerations (e.g., land-use and suitability for different feedstocks) and (2) changing temporal dynamics of supply and risk due to climate-related impacts on transportation networks (road, rail, pipeline), market international pricing volatility, and changes in policy targets and subsidy regimes as well as technological change associated with multi-generational biorefinery systems engineering advancements. Supply-chain models must continue to improve the ability to involve stochastic variables and multiobjective functions and constraints (cost, net GHG's, biodiversity, nutrient and water-efficiency, risk volatility) and to enable more flexible adjustment of model structure to become more user friendly (i.e., more transparent, visualization of model structure). This will, in turn, enable a broader base of input and involvement from a wide range of stakeholders to accelerate their development. For example, the same model, if designed with sufficient flexibility, could be applied to explore consequences and risks across a hierarchy of spatial applications (local, regional, trans-boundary). This would, in turn, aid in attaining more consistent and reliable information to examine sensitivity and evaluate uncertainty of predictions and forecasts across different spatiotemporal scales and application contexts.

ACKNOWLEDGMENTS

This paper was funded by the Canadian Federal Government via Agriculture and Agri-Food Canada's (AAFC's) Sustainable Agriculture Environmental Systems (SAGES) program. Historical crop yield data at the Consolidated Statistical Division (CSD) level was derived from Statistic Canada's Field Crop Reporting Series provided as an input to an AAFC research project (PI: Ted Huffman) on crop production and environmental modeling. The authors thank Mr. Zhirong Yang (AESB/AAFC) for assistance in generating spatial maps of frost kill and oilseed crop production across the Canadian Prairies.

REFERENCES

1.   P. Stelios and G. Mercier, "Renewable energy in Canada," Tech. Rep., Renewable Energy Working Party (REWP) of the International Energy Agency (IEA), Natural Resources Canada, Ottawa, Canada, 2002.

2.   J. Hill, E. Nelson, D. Tilman, S. Polasky, and D. Tiffany, "Environmental, economic, and energetic costs and benefits of biodiesel and ethanol biofuels," Proceedings of the National Academy of Sciences of the United States of America, vol. 103, no. 30, pp. 11206–11210, 2006.

3.   Environment Canada (EC), "Greenhouse gas sinks and sources in Canada, 1990–2009,"National Inventory Report, The Canadian Government's Submission to the United Nations Framework Convention on Climate Change (UNFCC), 2011,http://www.ec.gc.ca/ges-ghg/.

4.   N. Parker, P. Tittmann, Q. Hart et al., "Development of a biorefinery optimized biofuel supply curve for the Western United States," Biomass and Bioenergy, vol. 34, no. 11, pp. 1597–1607, 2010

5.   A. J. Dunnett, C. S. Adjiman, and N. Shah, "A spatially explicit whole-system model of the lignocellulosic bioethanol supply

chain: an assessment of decentralised processing potential," Biotechnology for Biofuels, vol. 1, p. 13, 2008.

6. X. van Tilberg, R. Egging, and M. Londo, Biofuel Production Chains: Background Document for Modeling the EU Biofuel Market Using the BioTrans Model, Energy Research Centre of the Netherlands, Policy Studies Unit, Petten, The Netherlands, 2005.

7. Y. Huang, C. W. Chen, and Y. Fan, "Multistage optimization of the supply chains of biofuels," Transportation Research E, vol. 46, no. 6, pp. 820–830, 2010.

8. S. Sokhansanj, A. Turhollow, and E. Wilkerson, "Development of the Integrated Biomass Supply Analysis and Modeling Tool (IBSAL)," Tech. Rep. ORNL/TM-2006/57, United States Department of Energy, Oak Ridge National Laboratory (ORNL), Environmental Sciences Division/Bioenergy Resources and Engineering Systems Group, 2008.

9. N. K. Newlands, J. Pinter, and B. McConkey, "Supply-chain optimization modeling of lingo-cellulosic ethanol within Canada: scoping document and model-development," Tech. Rep., Technical Report to the Canadian Federal Program of Energy Research and Development (PERD) and Canadian Biomass Innovation Network (CBIN), 2008, Agriculture and Agri-Food Canada (AAFC).

10. N. K. Newlands and L. Townley-Smith, Bayesian Network Model of Energy Crop Yield, Computational Intelligence, ACTA Press, Calgary, Canada, 2010.

11. Statistics Canada, "Cereals and oilseeds review," Tech. Rep. 22-007-X, Government of Canada, Ottawa, Canada, 2011.

12. N. K. Newlands, A. Davidson, H. Hill, and A. Howard, "Comparison of three methods for spatially interpolating daily precipitation and temperature across Canada,"Environmetrics, vol. 22, no. 2, pp. 205–223, 2010.

13. R. D. Elms and M. M. El-Halwagi, "The effect of greenhouse gas policy on the design and scheduling of biodiesel plants with multiple feedstocks," Clean Technologies and Environmental Policy, vol. 12, no. 5, pp. 547–560, 2010.

14. W. Zegada-Lizarazu and A. Monti, "Energy crops in rotation. A review," Biomass and Bioenergy, vol. 35, no. 1, pp. 12–25, 2011.

15. E. G. Smith, H. H. Janzen, and N. K. Newlands, "Energy balances of biodiesel production from soybean and canola in Canada," Canadian Journal of Plant Science, vol. 87, no. 4, pp. 793–801, 2007.

16. A. C. Kokossis and A. Yang, "On the use of systems technologies and a systematic approach for the synthesis and the design of future biorefineries," Computers and Chemical Engineering, vol. 34, no. 9, pp. 1397–1405, 2010.

17. L. Casseus, "Canola: a Canadian success story," Tech. Rep. 96-325-X, Government of Canada/Statistics Canada, Ottawa, Canada, 2009, Canadian Agriculture at a Glance.

18. R. E. Blackshaw, E. N. Johnson, Y. T. Gan, et al., "Alternative oilseed crops for biodiesel feedstock on the Canadian Prairies," Canadian Journal of Plant Science, vol. 91, no. 5, pp. 889–896, 2011.

19. A. Getinet, G. Rakow, and R. K. Downey, "Agronomic performance and seed quality of Ethiopian mustard in Saskatchewan," Canadian Journal of Plant Science, vol. 76, no. 3, pp. 387–392, 1996.

20. B. S. Rashford, C. T. Bastian, and J. G. Cole, "Agricultural land-use change in Prairie Canada: implications for wetland and waterfowl habitat conservation," Canadian Journal of Agricultural Economics, vol. 59, no. 2, pp. 185–205, 2011.

21. D. M. Yazan, A. Claudio Garavelli, A. Messeni Petruzzelli, and V. Albino, "The effect of spatial variables on the economic and environmental performance of bioenergy production chains," International Journal of Production Economics, vol. 131, no. 1, pp. 224–233, 2011.

22. B. Velazquez-Marti and E. Fernandez-Gonzalez, "Mathematical algorithms to locate factories to transform biomass in bioenergy focused on logistic network construction," Renewable Energy, vol. 35, no. 9, pp. 2136–2142, 2010.

23. J. Gan and C. T. Smith, "Optimal plant size and feedstock supply radius: a modeling approach to minimize bioenergy production costs," Biomass and Bioenergy, vol. 35, no. 8, pp. 3350–3359, 2011.

24. U. D. Tursen, S. Kang, H. Onal, Y. Ouyang, and J. Scheffran, "Optimal biorefinery locations and transportation network for the future biofuels industry in Illinois,"Resource Management and Policy, vol. 33, no. 3, pp. 151–173, 2010.

Chapter 7

Selection of Organisms for Systems Biology Study of Microbial Electricity Generation: A Review

Longfei Mao and Wynand S Verwoerd

Centre for Advanced Computational Solutions, Wine, Food and Molecular Bioscience Department, Lincoln University, Ellesmere Junction Road, Lincoln, 7647, New Zealand

ABSTRACT

A microbial fuel cell (MFC) is a device that uses microorganisms as biocatalysts to transform chemical energy or light energy into electricity. However, the commercial applications of MFCs are limited by their performance. This review presents the perspective that *in silico* metabolic modelling based on genome-scale metabolic networks can be used for understanding the metabolisms of the anodic microorganisms and optimizes the performance of their metabolic networks for MFCs. This is in contrast to conventional research that focuses on engineering designs and study of biological aspects of MFCs to improve interactions of anode and microorganisms. Four categories of biocatalysts - microalgae, cyanobacteria, geobacteria and yeast - are nominated for future *in silico* constraint-based modelling of MFCs after taking into account the cell type, operation mode, electron source and the availability of metabolic network specifications. In addition, the advantages and disadvantages of each organism for MFCs are discussed and compared.

REVIEW

Introduction

The technology for extracting an electrical current for use in external circuits from the metabolic processes of living microbes has been in development for more than a century [1]. The resulting devices, termed microbial fuel cells (MFCs), have several potential advantages over more prominent sustainable energy technologies such as solar or wind power. For example, they can directly convert organic waste into electricity [2] without pollution or inefficient intermediate steps that involve mechanical generators. This feature, energy recovery from solid wastes, has been exploited in proposed national strategies for many Asian countries [3]. It may be possible to achieve the same goal by inorganic catalysts or enzymes, but using living cells makes it possible to exploit their adaptability to environmental conditions and avoids the high capital cost of installation for other waste-to-energy systems reviewed by Eddine and Salah [4]. The whole organisms used in MFCs contain various enzymes and therefore allow different substrates (or mixed substrates) to be used. The organisms in the fuel cell system can be considered as micro-reactors and provide optimal conditions for different enzymes. Because the organisms are self-replicating, the organic matter oxidation implemented by these bio-catalysts is self-sustaining [5] and not subject to catalytic poisoning like metallic catalysts or degradation of enzyme catalysts. By selecting photosynthetic microbes, solar energy could be converted at the same time. One can envisage portable electronics powered by MFCs that are 'charged' by feeding them nutrients rather than electric current, or medical implants that derive their power directly from nutrients circulating in the bloodstream. Perhaps the process can be reversed, and external electrical power supplied to an MFC converted into biomass, as a temporary storage, to overcome the intermittent nature of many other sustainable energy sources - a possibility currently under serious consideration [6-8].

However, these future possibilities are still severely hampered by the low energy yields per mass or volume that are currently achieved. Generally MFC energy output is reported in milliwatts per square metre of electrode area or per cubic metre of electrolyte volume [9]. Scientific research has increased the densities of MFCs to over 1 kW m^{-3} (reactor volume) and to 6.9 W m^{-2} (anode area) under optimal conditions in the laboratory [10]. However, these values still cannot meet the needs of many applications, which require a power output larger than 100 kW m^{-2}[11]. For this reason large-scale waste water treatment is the application closest to industrial realisation and is the domain of much current MFC research.

A variety of designs are under development to improve the efficiency and potential application of MFCs to industry. Areas under investigation include the selection of electrode materials for optimal electrochemical performance and maximising electrode surface to volume ratios; improving charge transfer between microbes and electrodes either chemically or by mechanical design; and finding and maintaining optimal living conditions for microbe colonies, efficient supply of nutrients and removal of effluent. Different configurations are being investigated for extracting current, sometimes in combination with production of hydrogen or other metabolites of further use in energy generation and with or without exploitation of photosynthesis. The choice of process configuration and engineering design is also closely linked with the selection of the most suitable organism for a particular design or for whether overall priority is given to energy generation, waste disposal or some other objective.

A schematic representation of MFC research activity is shown in Figure 1. While there is a large volume of biochemical research literature on, for example, electron transfer chains and redox processes in cell metabolism that is relevant to MFC, relatively few studies focus specifically on MFC. This is exacerbated by the fact that ongoing research continues to identify new mechanisms for electron exchange between microbes and electrodes, new design strategies to exploit these and consequently new candidate organisms. Such organisms have not necessarily been well studied experimentally before.

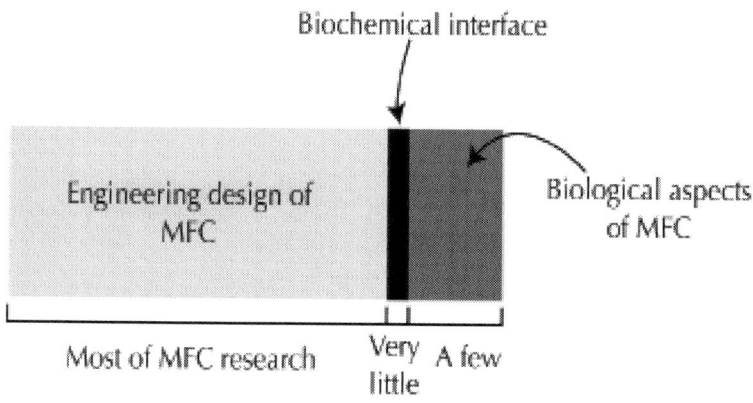

Figure 1: Areas covered in current microbial fuel cell research.

In silico modelling is well suited to bridge this gap and extend knowledge in the biochemical interface between MFC biology and engineering design. Externally, electron flow (in the external circuit) and the counterflow of protons in the electrolyte make up the current that carries useful electrical power. Internally, both of these are comprehensively woven into the fabric of metabolism: electrons being transported by redox carrier molecules such as reduced nicotinamide adenine dinucleotide (NADH) that participate in a large fraction of all biochemical reactions, and protons that, for example, drive adenosine triphosphate (ATP) synthesis needed for energy transport are also ubiquitously involved in a great many reactions. This clearly calls for a systems-level approach rather than the reductionist strategy of pathway-oriented, conventional biochemistry.

This is the domain of systems biology [12]. Systems biology provides *in silico* models that incorporate biological data, metabolic flux data and different physico-chemical constraints such as the conservation of mass and energy, thermodynamics, redox balance, etc. [12] and thus provides an opportunity to identify the bottlenecks hidden in a complex network of interactions and cellular compartmentation [13].

The kinetic behaviour of a metabolic network at a whole-genome level can be constructed and analysed through a mathematical model [14,15]. However, the characterization of metabolic networks is still far from comprehensive in databases [16] and even in the best-understood organisms, the majority of kinetic parameters are undetermined.

The development of new computational methods allows for the whole-network modelling of metabolism and conduction of compelling and testable predictions even without many parameters. The key idea is to incorporate stoichiometry and other fundamental principles as mathematical constraints, which separate feasible and infeasible metabolic behaviours. Compared with kinetic parameters, these constraints are much easier to identify and make it possible to build a large-scale model [17]. These constraint-based modelling approaches allow integration of high-throughput post-genomic data but describe steady states and generally offer no information about metabolite concentrations or the temporal dynamics of the system [18-20].

Genome-scale metabolic modelling requires high-quality metabolic network reconstructions [14]. The reconstructions are based on sequenced genomes and are generally built manually using information from metabolic databases (e.g. KEGG and BRENDA) and primary literature. The metabolic network reconstruction process is described in detail elsewhere [21]. Recently, due to the development of the high-throughput technologies, the reconstructions have now been built for various organisms [22].

Flux analysis can be combined with cell biology and sub-cellular biochemistry to reveal the functionality and efficiency of the enzymes associated with cell biological components or structures[23]. Metabolic regulation can be deeply understood only when multiple system components are examined simultaneously. This kind of analysis has been conducted in microbial and medical research in recent years.

Flux determinations can produce results that are hardly predictable from observed changes in transcript or protein levels because most of metabolic control takes place at post-translational levels and enzyme activities are often not correlated with changes in transcript or protein levels [23]. The incorporation of data from enzyme platforms should make the functionality of genomics strategies more clear. System-wide metabolic flux characterization is an important part of metabolic engineering [23].

Nevertheless, there is no published literature that uses genome-scale flux models to study the metabolic behaviour of biocatalysts in MFCs. The only attempt to use a genome-scale model to study biocatalyst behaviour in an MFC was presented as a conference abstract paper in the 17th European Symposium on Computer Aided Process Engineering (2007) [24]. However, this paper is an immature work that did not provide the source of the central metabolic network, describe the gene knockout methodology or discuss the results in relation to other MFC experimental work. Therefore, future research activities are urgently needed to fill the research gap indicated in Figure 1, with recently advanced constraint-based modelling approaches.

An essential first step in applying constraint-based analysis to microbial fuel cells is to choose appropriate organisms for further study. Due to the varied strategies and designs alluded to above, no single organism can serve as a suitable model, and a major advantage of *in silico* modelling is that different organisms can be studied in the same framework to facilitate mutual comparisons. This paper reviews the background against which such choices can be made and proposes a set of four organisms for the purpose.

The 'Microbial fuel cells' section explains the construction, operation and classification of current MFC designs, and the 'Current directions of MFC research' section reviews the issues being addressed in current MFC research. Based on this, the 'Microorganisms for *in silico* study of MFC functioning' section discusses a selection of organisms that are representative of various combinations of biological aspects that can be exploited in MFCs, while also featuring well-established metabolic network reconstructions, suitable for the computational analysis.

Microbial Fuel Cells

MFCs are unique devices that can use microorganisms as catalysts for transforming chemical energy directly into electricity. The biggest advantage of an MFC is that it can generate combustion-less, pollution-free bioelectricity directly from the organic matter in biomass [2]. In an MFC the energy stored in chemical bonds in organic compounds is converted to electrical energy through enzymatic reactions by microorganisms. Thus, the electricity production by MFC is associated with the normal living processes of bacteria capturing and processing energy.

In a typical MFC configuration (Figure 2), microorganisms are situated in the anodic compartment and use the biomass for growth while forming electrons and protons [25]. The electrons are transported out of cells to an electrode using redox mediators or directly expelled by some microorganisms for reducing the substrate. The protons or H+ ions are diffused through the electrolyte to the cathode where it is oxidised to water. The cathode can be in a separate chamber (i.e. double-chambered MFCs) or in the same chamber (i.e. single-chambered MFCs). A single-chambered MFC eliminates the need for the cathodic chamber by exposing the cathode directly to the air. The only by-product released by MFCs is carbon dioxide, which can be fixed by plants for photosynthesis.

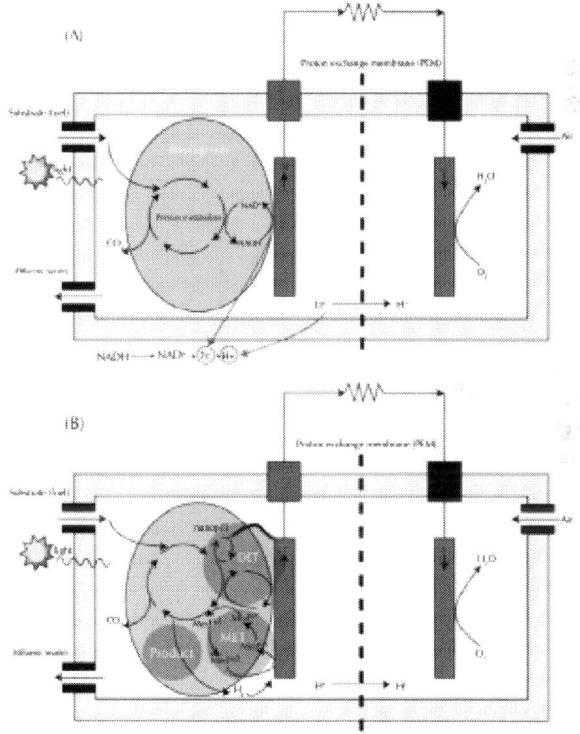

Figure 2: The working principle of a microbial fuel cell. (A) A bacterium in the anode compartment transfers electrons obtained from an electron donor (glucose or light in the case of photosynthetic organisms) to the anode electrode. Protons are also produced in excess during electron production.

These protons flow through the proton exchange membrane (PEM) into the cathode chamber. The electrons flow from the anode through an external resistance (or load) to the cathode where they react with the final electron acceptor (oxygen) and protons. (B) Three electron transfer modes: (1) directly via membrane-associated components (DET), (2) mediated by soluble electron shuttles (MET) or (3) primary product (Product), e.g. H_2 can act as a fuel to be oxidised to provide electrons for the electricity circuit. Med, redox mediator; Red oval, terminal electron shuttle in or on the bacterium.

MFCs require running under conditions predefined by the optimum growth and living conditions of the used microorganisms. Thus, factors affecting the MFC's efficiency include electrode material, pH buffer and electrolyte, proton exchange system and operating conditions in both the anodic chamber and the cathodic chamber. MFCs are usually operated at ambient temperature, at atmospheric pressure and at pH conditions that are neutral or only slightly acidic [26].

MFCs harness the electrons from these systems in three main operation modes: mediated electron transfer (MET), direct electron transfer (DET) and product mode. Photosynthetic MFCs use photosynthesis as the electron source and can also be operated in the same modes.

Mediated Electron Transfer

MET is defined as where a mediator molecule acts an electron relay that repeatedly cycles between the reaction sites and the electrode [11]. MET is the most common electron transfer mode used in MFCs and can be classified into two sub-types [11]:

- Indirect transfer systems that involve freely diffusing mediator molecules (i.e. diffusive MET)

- Indirect transfer systems in which the mediator is integrated into the electrode or the cell membrane (i.e. non-diffusive MET)

In diffusive MET, the mediators enter the cell membrane and exchange electrons between cellular metabolism inside the cell and the electrode outside it. In the non-diffusive MET, the mediator can collect the electrons from the cell membrane without penetrating the cell.

Based on the type of mediators, diffusive MET can also be classified into three sub-categories:

- MET via exogenous (artificial) redox mediators
- MET via secondary metabolites
- MET via primary metabolites

The detailed mechanisms in those three classes are discussed in [27]. Because the terminal electron transfer to or from the electrode determines the overall cell potential, potential (voltage) losses can be minimised by using a mediator that has a reaction potential near that of the biological component.

Direct Electron Transfer

DET is defined as the case where electrons cycle directly between a microorganism and an electrode. DET can be achieved through two naturally occurring mechanisms:

- Membrane-bound c-type cytochromes, which exist in the cell membrane in some organisms [28,29] to provide electron transfer capacity. For example, multi-heme proteins have especially evolved in sediment-inhabiting metal-reducing microorganisms such as *Geobacter*[30],*Rhodoferax*[31] and *Shewanella*[32]. In their natural environment, iron (III) oxides act as the solid terminal electron acceptors, but in the case of MFC, the anode is used as the solid electron acceptor.

- Electronically conducting nanowires. The DET via outer membrane cytochromes requires the cytochrome (the bacterial cell) to be physically adhered to the fuel cell anode. When a biofilm is formed, only bacteria in the first monolayer at the anode surface are electrochemically active [30]. Thus, the maximum cell density in this bacterial monolayer usually influences the MFC performance. However, it has been shown that some *Geobacter* and the *Shewanella* strains can evolve electronically conducting molecular pili (nanowires of 2 to 3 μm long, made of fibrous protein structures [33]) that make the microorganism able to reach and use more distant solid electron acceptors [34,35]. The pili are connected to the membrane-bound cytochromes and allow transference of the electrons to the distant electron acceptors without cellular contact (Figure 2B). Thus, thicker electroactive biofilms can be formed to increase anode performance. It was shown that fuel cell performance can be increased up to tenfold upon nanowire formation of *Geobacter sulfurreducens* [35].

Product Type

In product-type MFCs, microbes metabolize the substrate, releasing a secondary fuel product such as hydrogen that then diffuses to the electrode and is oxidised or reduced (as appropriate) to form a final waste product, which is discharged [11]. The product operation is similar to conventional fermentation processes, in that products of the microbial metabolism are used as the fuel at the electrode.

The product system is made up of two independent stages: one is storage of the microbial reaction product, and the other one is the product being fed to a conventional fuel cell process driven by non-biological catalysis, such as in the case of a proton exchange membrane fuel cell, where H_2 is converted into electricity [36]. These stages may also be physically separated in different containment vessels. However, a product system only truly belongs to a biofuel cell system when the microbes and the electrode are together in the same anode compartment [11]. Nowadays, the fermentation (mostly to hydrogen gas) usually takes place in the fuel cell itself [37,38].

Product systems have two main drawbacks, one is that the efficiency of the conversion of the biological substrate to hydrogen is quite low, and the other is that hydrogen oxidation requires high fuel cell temperatures. Also, the produced biofuel gas is always contaminated with other by-products such as CO, H_2S and (poly)siloxanes making it not sufficiently pure for direct use in a fuel cell [39].

Photosynthetic MFCs

Photosynthetic MFCs are MFCs that generate electricity from a light source rather than a fuel substrate and require the mediator involved to be light stable [40]. Conventionally two operating modes exist for photosynthetic MFCs:

- Energy is produced and stored by the microorganism during illumination and then released and processed in the same way as in a non-photosynthetic biofuel cell.

- The energy produced during illumination may be directly extracted in the form of electrons for creating an external electrical circuit.

A single photosynthetic MFC may possess both of these two modes of action. However, it is recently thought better to classify photosynthetic MFCs into categories based on seven approaches that integrate photosynthesis with MFCs - photosynthetic MFCs [40]:

- Photosynthetic bacteria at the anode with artificial mediators
- Hydrogen-generating photosynthetic bacteria with an electrocatalytic anode
- A mixed culture, with photosynthetic bacteria supplying organic matter to heterotrophic electroactive bacteria at the anode
- Photosynthesis in plants, supplying organic matter via rhizodeposits to heterotrophic electroactive bacteria at the anode
- An external photosynthetic bioreactor, where only biomass or metabolic products are transferred to the anode compartment to feed heterotrophic electroactive bacteria
- Direct electron transfer between photosynthetic bacteria and electrodes
- Photosynthesis at the cathode to provide oxygen

These sub-types have been discussed in detail by Rosenbaum et al. [40].

Microorganisms Suitable for MFCs

Most microorganisms are unable to donate sufficient electrons outside of cells to produce usable currents, because the outer layers of most microbial species are made up of non-conductive lipid membrane, peptididoglycans and lipopolysaccharides which restrain electron transfer to the anode [41]. Since the 1980s, it has been found that artificial water-soluble electron shuttles (i.e. methylene blue, thionine, neutral red and 2-hydroxy-1,4-naphthoquinone) can be used as mediators that transport the electrons from electron carrier molecules inside the cell (e.g. NADH, NADPH or reduced cytochromes) to the anode surface [41]. For example, an MFC based on *Proteus vulgaris* used thionine as a mediator to generate electricity from sucrose [42].

Since the 1990s, some bacterial species such as *Pseudomonas aeruginosa* [43] and *Clostridium butyricumcan*[44] have been found to be able to self-mediate extracellular electron transfer using their own metabolic products. Meanwhile, direct transfer of electrons (DET) that involves use of electrochemically active redox enzymes (i.e. cytochromes) has been discovered in a number of bacterial species such as *Shewanella putrefaciens*[28,29,45], *Shewanella oneidensis*[46], *Geobacter sulfurreducens*[a][30,47], *Rhodoferax ferrireducens*[31], and the oxygenic phototrophic cyanobacterium *Synechocystis* sp. PCC 6803[a][34]. These microorganisms are termed as exoelectrogens, and among them *S. oneidensis* and *G. sulfurreducens* have evolved electronically conducting molecular pili (nanowires) to further facilitate the DET [34,35]. Besides DET mode, *S. oneidensis* can conduct MET using a self-produced mediator [48]. The exoelectrogens in MFCs are thought to actively use electrodes to conserve electrochemical energy required for their growth and thus ensure high rates of fuel oxidation and electron transfer for the production of electrical energy [5].

In most of the previous MFC studies, bacteria have been used for electricity generation. On the other hand, since 2000, eukaryotes such as microalgae (e.g. *Chlamydomonas reinhardtii*[a]) and yeast (e.g. *Saccharomyces cerevisiae*[a]) have also emerged as good choices for MFC use, because they have been studied as model microorganisms in the lab and have been widely used in the industry for a long time.

[a]Representative microorganisms chosen by this article and discussed in the 'Microorganisms for *in silico* study of MFC functioning' section.

Current Directions of MFC Research

Engineering Design and Biological Aspects

Most previous studies tended to improve power densities of MFCs by optimizing the reactor configuration and operation parameters [49,50], such as modifications of the electrode materials to incorporate metals that contain current collectors [51,52], use of metals highly optimized for bacterial adhesion and metals possessing high electrical conductivity to minimise ohmic losses[10], and application of a biocathode that can increase MFC performance by improved oxidation of hydrogen at the cathode [53]. Applications of chemical treatments and precious metals to electrodes in order to increase power production in the laboratory have also been investigated [54,55]. However, these modifications, like the use of larger laboratory reactors, may increase the cost and lead to compromises on performance based on material costs. Many bottlenecks also exist for improving those physical and chemical properties.

Since the fundamental source of electrons is the cellular metabolism, it is particularly important to focus on biological processes that take place in the microbial cells. In this regard, further clarification is needed of the factors relevant to the anodic catalysis process, such as the diversity of the electrochemically active microorganisms [56] and especially the electricity generation mechanisms in relation to normal metabolic states.

Mediator-less, Mediator-self-producing and Artificial Mediator-based MFCs

Mediator-less MFCs are a more recent development relying on evolved ability of exoelectrogens for disposal of electrons originating from substrate oxidation. This type of MFCs has been increasingly preferred, because use of mediators complicates the cell design and these mediators are usually toxic, costly and unsustainable, limiting MFC development [27,57-60]. In addition, mediator-involved MFCs usually produce low current densities (0.1 to 1 A m^{-2}) [27]. Unfortunately, mediator-less MFCs may not yet find a wide range of applications since the discovered exoelectrogens are still few in number and it is non-exoelectrogens that are largely used in the agricultural and industrial areas [61]. Thus, an important direction of MFC research is development of MFCs using non-exoelectrogens without exogenous mediators [61].

However, problems arise from the fact that redox molecules used in electron transfer reactions are not situated on the outer membrane, but in the cytoplasmic membrane. One way is to develop direct electron transfer using carbon nanoparticles that can contact the redox centres that are incorporated in the interior cell membrane [61]. Another way is to identify and develop self-produced mediators (e.g. in the case of *Shewanella* species mentioned above) through engineering methods [56].

Conventional Photosynthetic MFCs

It is appealing to study whether phototrophic microbes can be used in an MFC generating electricity from sunlight, because sunlight is an unlimited energy resource and more solar energy reaches the Earth in 1 h (4.1×10^{20} J) than the energy consumed on the planet in a year[56,62,63]. In addition, the development of a self-sustainable photosynthetic MFC is important to meeting energy requirements at remote locations, where routine addition of fuel would be technically difficult and expensive [56].

Photosynthetic MFCs can generate electricity indirectly or directly. For example, in the indirect way, *Rhodobacter sphaeroides* in the MFC can produce H_2 that is oxidised at a platinum coated anode to generate electricity [64,65], whereas in the direct way, *Rhodopseudomonas palustris*, a photosynthetic purple non-sulphur bacterium, can generate electricity in a biofilm anode MFC by direct electron transfer [66].

There are also more traditional photosynthetic MFC configurations where photosynthetic organisms live with other microbes and supply products to heterotrophs [67]. Photosynthetic microorganisms (e.g. cyanobacteria or microalgae) and heterotrophic bacteria exhibit synergistic interactions [68] that can be used in self-sustained phototrophic MFCs [62]. An indirect synergistic relationship between photosynthetic organisms and electricigens has been exemplified in a recent study, in which algal photobioreactors were used to supply organic matter produced via photosynthesis to an MFC for electricity generation [69]. The operation of this type of photosynthetic MFCs is CO_2 neutral and does not need buffers or exogenous electron transfer mediators [67]. However, the photosynthetic MFC power densities obtained are quite low when compared with those that are currently reported for conventional MFCs, e.g. 0.95 mW m^{-2} for polyaniline-coated and 1.3 mW m^{-2} for polypyrrole-coated anodes [70] versus values in the watt per square metre range for conventional cells.

MFCs Based on the Photosynthetic Electron Transfer Chain

Recently, the photosynthetic electron transfer chain is considered as a source of the electrons harvested on the anode surface, which is different from the previously designed anaerobic MFCs, sediment MFCs or anaerobic photosynthetic MFCs [70]. A single-chamber photosynthetic MFC based on two photosynthetic cultures, planktonic cyanobacteria *Synechocystis* sp. PCC 6803 and a natural freshwater biofilm, has shown a positive light response (i.e. immediate increase in current upon illumination) [70]. This phenomenon proves that it is possible to extract electrons directly from the photosynthetic electron transfer chain, and not only from the respiratory transfer chain or through oxidation of hydrogen [71].

Microorganisms for *in silico* Study of MFC Functioning

Categories and Representatives

Because the electricity generation in MFCs is based on the metabolic activity of living microorganisms, experimental screening of different microorganisms for better anodic activity has long been recognised as a fundamental way to improve MFC performance. It is also possible to improve the performance by culturing microorganisms under selective pressure for enhanced power production [30,72].

Compared to experimental studies, *in silico* modelling is less constrained to a particular MFC design and operating mode. Clearly no single organism is likely to be optimal for all of the varied designs discussed before. To date, every microorganism used in previous MFC studies has advantages and disadvantages. Selection of the microorganism depends on a variety of factors such as types of application, the capability of power generation, the availability of types of energy source for bacterial survival and the ability of extracellular electron transfer, in that electrodes are not natural electron acceptors.

From the modelling perspective, a broader view is possible. Categories of microbes can be identified to cover the range of operating modes and, within these, individual organisms selected that will allow different modes to be individually studied and also compared quantitatively.

MFC microbial communities can be divided into three groups: heterotrophic cells, photoheterotrophic cells and sedimentary cells [9].

The distinction between phototrophic and heterotrophic metabolism is fundamentally important in determining the operating mode. Another key distinction is between prokaryotes and eukaryotes. Compared to prokaryotic species and mixed cultures that have been mostly studied for different MFC applications, fewer studies involve eukaryotes as biocatalysts in MFC operations [73]. This is because the metabolic processes of eukaryotic cells take place in the membrane surrounded cell organelles (e.g. chloroplasts) and is thus putative to be difficult for some commonly used redox mediators such as 2-hydroxy-1,4-nepthoquinone to get access to [57,74]. While prokaryotes have the advantage that their simpler cell membranes and internal structure are more amenable for physical electron extraction, the more complex metabolism of eukaryotes may be more efficient and be able to support a larger diversion of redox carrier flux without undue harmful effects on the organism.

The four anodic microorganisms: C. reinhardtii, Synechocystis sp. PCC 6803, S. cerevisiae and G. sulfurreducens, each combine a different pair of key features and are proposed as good candidates for MFC *in silico* characterization. As illustrated in Figure 3, C. reinhardtii and S. cerevisiae are eukaryotes, whereas Synechocystis sp. PCC 6803 and G. sulfurreducens are prokaryotes. The four organisms also cover the three groups of the MFC microbial communities mentioned above, i.e. C. reinhardtii and Synechocystis sp. PCC 6803 are photoheterotrophic cells, S. cerevisiae belongs to heterotrophic cells and G. sulfurreducens is a typical sedimentary cell.

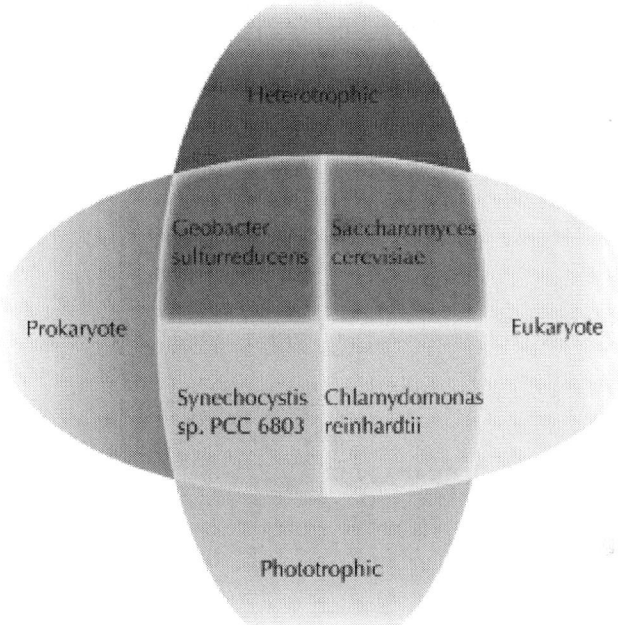

Figure 3: Classification of proposed microbes

C. reinhardtii and *Synechocystis* sp. PCC 6803 are photosynthetic organisms that are also capable of producing hydrogen. The comparison of organisms with/without photosynthesis can be used to study exploitation of photosynthetic and respiratory electron transport chains to supply MFC current.

A further consideration in a modelling study is whether the data are available and computationally manageable. All four microorganisms have been studied extensively as model organisms and used in various industries for a long time, and thus, related molecular tools and biological mechanisms are abundant. In particular, genome-scale metabolic networks have been reconstructed and are regularly updated for these four organisms. Based on a literature review, the most updated models are shown in Table 1. These models include the natural redox mediators (i.e. NADH) that are well balanced for the cellular energy metabolisms (e.g. oxidative phosphorylation, glycolysis, Calvin cycles and tricarboxylic acid (TCA) cycles) and are thus practicable for MFC modelling.

Table 1: The scope of the genome-scale models of the four selected organisms

	Chlamydomonas reinhardtii	Synechocystissp. PCC 6083		Geobacter sulfurreducens	Saccharomyces cerevisiae
Gene	2,249	1,080	1,811	678	617	918
Metabolite	1,862	1,068	465	795	644	1,655
Reaction	1,725	2,190	493	863	709	2,110
Compartment	4	10	3	4	2	17
Date	December 2011	August 2011	October 2011	January 2012	2009	2012
Reference	[75]	[76]	[77]	[78]	[79]	[80]

The number indicates the counts of relative items in the network models.

Mao and Verwoerd

Mao and Verwoerd International Journal of Energy and Environmental Engineering 2013 4:17, doi:10.1186/2251-6832-4-17

It is noted that two models were published in the same period for *C. reinhardtii*[75,76] and*Synechocystis*[77,78]. These models for the same organisms are not compared with each other by the authors, and thus, their limitations can only be revealed during the MFC modelling.

The biological features of the four microorganisms are summarized in Table 2, and the relevance of each of these microbes is reviewed in more detail in the following sections.

Table 2: Comparison of facts of four selected microorganisms

Name	*Chlamydomonas reinhardtii*	*Synechocystis sp. PCC 6083*	*Saccharomyces cerevisiae*	*Geobacter sulfurreducens*
Domain	Eukaryote	Prokaryote (Gram-negative)	Eukaryote	Prokaryote (Gram-negative)
Mitochondria	Multiple	N/A	Multiple	N/A
Chloroplast	Single chloroplast occupies two thirds of the cell	Chloroplast analogy	N/A	N/A
Hydrogen synthesis enzyme	Fe hydrogenase	NiFe hydrogenase	N/A	N/A
MFC mode	Product mode	Photosynthetic DET	MET DET (output extremely low)	DET
MFC performance	0.4 W m^{-2}	0.02 W m^{-2}	1.5 W m^{-2}(MET)	1.88 W m^{-2}
	3.3 W m^{-3}	0.007 W m^{-3}	90 W m^{-3}(MET)	43 W m^{-3}
			0.003 W m^{-2} (DET)	
Optimum growth temperature	20°C to 25°C [81]	30°C to 33°C [82]	25°C to 35°C [83]	30°C to 35°C [84]
Growth mode	Autotrophic	Autotrophic	Heterotrophic	Heterotrophic and sedimentary (soil inhabitant)
	Heterotrophic	Heterotrophic		
	Mixotrophic	Mixotrophic		

N/A, not applicable

Mao and Verwoerd

Mao and Verwoerd *International Journal of Energy and Environmental Engineering* 2013 4:17, doi:10.1186/2251-6832-4-17

Chlamydomonas Reinhardtii

C. reinhardtii is a unicellular green alga that belongs to the Chlorophytes division, which diverged from the Streptophytes division (including land plants) more than one billion years ago [85]. *C. reinhardtii* is an approximately 10-µm, unicellular, soil-dwelling green alga with an eyespot, a nucleus, multiple mitochondria, two anterior flagella for motility and mating, and a single cup-shaped chloroplast that accommodates the photosynthetic apparatus [86,87].

Like plants, *C. reinhardtii* has a cell wall and can grow in a medium lacking carbon and energy sources when illuminated [87]. Unlike angiosperms (flowering plants), this microorganism has functional photosynthetic apparatus even when in dark conditions and with an organic carbon source [87]. In the dark, acetate is the sole carbon source used by wild-type *C. reinhardtii in vivo*[88].

Because of the relative adaptability and quick generation time, *C. reinhardtii* has been used as a model to study eukaryotic photosynthesis, eukaryotic flagella and basal body functions and the pathological effects of their dysfunction [89,90], and investigated for water bioremediation and biofuel generation [87,91-93]. The cDNA, genomic sequence and mutant strains of *C. reinhardtii*are publicly available through the *Chlamydomonas* Center [94].

Advantages of Algae

C. reinhardtii inherits all potential advantages of algae for industrial use and scientific study, including [95-97] the following: (1) Algae biomass can potentially be produced at extremely high volumes, and this biomass can yield a much higher oil (1,000 to 4,000 gal/acre/year) than soybeans and other oil crops [97]. (2) Algae do not compete with traditional agriculture because they are a non-food-based resource which can be cultivated in large open ponds or in closed photobioreactors located on low-productive or non-arable land. (3) Algae have a good adaption to different climate and water conditions and can be grown in a wide range of water sources, such as brackish, saline, or fresh and waste water. (4) Algae can make use of resources that would otherwise be considered waste as substrate for growth [97]. (5) Algae can use and sequester CO_2 from many sources such as flue gases of fossil fuel power plants and other waste streams. (6) Algae can be processed into a wide range of products such as biodiesel, bioethanol, methane, bio-oil and biochar, and high-protein animal feed. (7) The 'simple' photosynthetic alga *C. reinhardtii* is an excellent model organism for a systems biology approach compared to a complex vascular plant [12].

Biofuel and Electricity Production by Algae

Due to the advantages listed above, algae have been examined in many studies for the generation of energy products, such as bio-oil, methane, methanol and hydrogen [98]. Nevertheless, these technologies have one disadvantage in that the fuel produced must be stored, transported and further processed to produce electricity. To circumvent these problems, MFC is used as an alternative way to directly generate electricity in only one process unit by means of hydrolysis and fermentation of algae and makes use of energy originating from sunlight.

However, algae are not exoelectrogens and the conventional mediators do not perform well in the extraction of the redox potential for algae-based MFCs because the redox species are produced through the metabolic mechanisms that take place in membrane-surrounded cell organelles in algae such as mitochondria and chloroplasts [74]. Thus, neither DET nor MET mode has been applied in MFCs using algae.

Previous studies tend to use algae in MFCs of the product mode that depend on the production of hydrogen molecules, which is then oxidised at the anode for electron transfer to the MFC circuit. Another mechanism is where algae produce organic matter that is used as a substrate for electrochemically active bacteria, which then supply electrons for MFCs from oxidization of the organic matter [69].

In one landmark study of MFC using algae, *C. reinhardtii* was used in a product-mode MFC to produce hydrogen for oxidation at the anode. A maximum hydrogen production rate of 7.6 ml/l culture h^{-1} [99] was achieved, at a current yield of 9 mA at a constant electrode potential of 0.2 V. Using the culture volume and electrode dimensions, this corresponds to power densities of 0.4 W m^{-2} and 3.3 W m^{-3}. In another study [98], *Chlorella vulgaris* microalgae were used as a biomass source to feed a mixed microbial culture, producing a maximum power density of 0.98 W m^{-2} (277 W m^{-3}).

Furthermore, algae have been inoculated into a bioelectrode to generate oxygen as the electron acceptor [100]. Under illumination, algae produced oxygen as the electron acceptor for the MFC cathodic reactions, changing the bioelectrode into biocathode mode, while in darkness, the algal oxygen production stops and the bioelectrode mainly functioned as the bioanode. The reversible bioelectrode can relieve the pH membrane gradient generated by the acidification at the anode and the alkalisation at the cathode during normal MFC operation [100,101].

Hydrogen Production by C. Reinhardtii

Only a specific group of green microalgae and cyanobacteria, e.g. microalga *C. reinhardtii*, have evolved the additional ability to harness the huge solar energy resource to drive molecular H_2 production [102-108]. The release of hydrogen by *C. reinhardtii* under light exposure was first reported in 1942 [105]. In 2000, sustained hydrogen production was achieved using induced sulphur depletion in a culture medium containing acetate, a carbon source that is used to cause the shift from aerobic to anaerobic state [107].

C. reinhardtii is one of the best eukaryotes for H_2 production [109]. The available experimental information, including genomics, indicates that *C. reinhardtii* possesses a complex metabolic network containing aerobic respiration and molecular flexibility associated with fermentative metabolism. The molecular flexibility is accomplished with adjustments in the rates of accumulation of organic acids, ethanol, CO_2 and H_2 [86,110-115] and underlies the cell's adaptive ability for hypoxic and anoxic conditions.

Compared to other H_2-producing organisms such as chemotrophic and phototrophic bacteria, *C. reinhardtii* is more practical for H_2 production as it can be easily and efficiently grown in bioreactors using solar light, grows rapidly (doubling times of the order 6 h or less) and has a flexible metabolism [116]. The genome of this model microorganism was fully sequenced in 2007 [86], which makes it possible to increase production yields of H_2 from water by optimization of cell metabolism.

Limitation of Hydrogen Production by C. Reinhardtii

In fact, hydrogen production by C. reinhardtii can still not meet the commercial requirement because of several biochemical and engineering shortcomings, for example, hydrogen production demands anoxia because oxygen can suppress transcription and function of hydrogenase(s). However, the anoxia is constrained by the function of the photosystem II (PSII), which provides electron and protons from water and conducts oxygen evolution in the photosynthetic electron transport chain. Economic assessments have suggested that microalgae should achieve an efficiency of 10% in the conversion of solar energy to bioenergy to be competitive with other H_2 production methods, such as biomass gasification or photovoltaic electrolysis [117]. This is a more than a fivefold increase in efficiency from current levels. Exploiting hydrogen production directly in a product-type MFC may help to bridge this gap.

Synechocystis sp. PCC 6803

Synechocystis sp. PCC 6803 is a unicellular cyanobacterium, one of the earliest groups of microbes to evolve on earth. The first primitive bacteria on Earth are dated at 3.8 to 3.6 billion years ago [118]. It is thought that cyanobacteria flourished during the period from 3.5 to 1.8 billion years ago, consuming CO_2 and providing Earth with oxygen, making possible the development of the different forms of aerobic life. At present, cyanobacteria deliver amounts of oxygen to the atmosphere similar to those that are produced by higher plants [118]. Moreover, cyanobacteria harness 0.2% to 0.3% of the total solar energy (178,000 TW) that reaches the Earth [106] and convert the solar energy into biomass-stored chemical energy at the rate of approximately 450 TW, contributing to 20% to 30% of Earth's primary photosynthetic productivity [119].

Until 1982, the cyanobacteria were called blue-green algae because they can photosynthesize and look like chloroplasts. Since then, cyanobacteria were re-classified as prokaryotes [120]. It is suggested that cyanobacteria entered into a symbiosis with cells, which were not capable of absorbing CO_2 and releasing oxygen, and later became photosynthetic organelles of plants [121]. Nowadays many species of cyanobacteria, e.g. *Synechocystis* sp. PCC 6803, are widely distributed in nature.

Synechocystis sp. PCC 6803 and plants have similar oxygen-evolving apparatus and are thus used for studying photosynthesis in plant cells. The difference is that *Synechocystis* sp. PCC 6803 grow much faster than plants, and they are relatively easy organisms for genetic manipulations [118]. Also, plants are fixed at places where they grow and they have less adaptation abilities for their growth and propagation than cyanobacteria.

Synechocystis sp. PCC 6803 grow photoautotrophically on carbon dioxide and light, as well as heterotrophically on glucose. Like *C. reinhardtii*, *Synechocystis* sp. PCC 6803 is one of several hydrogen-yielding species of cyanobacteria [122]. After its genome was fully sequenced in the 1990s [123,124], this cyanobacteria species has become a popular model photosynthetic organism studied by many researchers.

Advantages of Cyanobacteria

Cyanobacteria, besides other photosynthetic microorganisms such as microalgae, can establish synergistic relationships with heterotrophic bacteria, for instance, in a microbial mat [68]. Thus, they could be potentially manipulated to establish an indirect synergistic relationship with electricigens in phototrophic MFC [69]. However, phototrophic MFCs usually have low conversion efficiency [62], and the study of phototrophic MFCs is in its nascent stages [62].

Since *Synechocystis* sp. PCC 6803 is a photoautotroph that divides rapidly, it has been enlisted as a platform for production of biofuels by using sunlight as an inexpensive energy source [125,126]. This feature makes this species suitable as a candidate for MFCs of product mode.

Electrogenic Activity of Cyanobacteria

Unlike other exoelectrogens, such as *G. sulfurreducens*, in which the electrons are derived from biochemical oxidation of organic compounds via the respiratory electron transfer chain [127], cyanobacterial electrogenic activity does not need exogenous organic fuel and is entirely dependent on the energy of light, which drives the biophotolysis of water through the photosynthetic electron transfer chain in the cyanobacteria, releasing electrons [128,129]. The electrogenic activity of cyanobacteria may represent a form of overflow metabolism to protect cells under high-intensity light [70,129]. This light-driven electrogenic activity is conserved in diverse genera of cyanobacteria and is an important microbiological channel of solar energy into the biosphere [129].

The electrogenic activity of *Synechocystis* sp. PCC 6803 has been captured in an MFC for electricity generation. The MFC can achieve a steady power density of $6.7 \, mW \, m^{-3}$ (peaking at $7.5 \, mW \, m^{-3}$) [130,131]. These power densities are still much lower than the values achieved by the other microbes under discussion. Despite that, it is included in the selection list because it offers a unique combination of photosynthetic activity that is plausibly accessible to direct-mode electron transfer. The quoted measurements are quite recent, and it is worth exploring if this organism has the potential to deliver competitive power densities in the future.

Hydrogen Production by Cyanobacteria

Cyanobacteria have a similar process for hydrogen production as algae, except that they use NiFe hydrogenases rather than Fe hydrogenases in microalgae and the hydrogenase of cyanobacteria is 100 times less active than those of the green algae *C. reinhardtii*[106]. These hydrogenases contain [Ni-Fe] catalytic centres that are extremely sensitive to inactivation by O_2, one of the major barriers to hydrogen production. Natural mechanisms such as consumption by respiration, chemical reduction via PSI and reversible inactivation of PSII O_2 evolution can reduce intracellular O_2 content and thereby increase H_2 production.

Saccharomyces Cerevisiae

The *Saccharomyces* genus currently contains eight species [132]. *Saccharomyces cerevisiae*, *Saccharomyces bayanus* and *Saccharomyces pastorianus* are associated with anthropic environments, whereas *Saccharomyces   paradoxus,   Saccharomyces   kudriavzevii*,

Saccharomyces cariocanus, *Saccharomyces mikatae* and the recently described *Saccharomyces arboricolus* are mostly isolated from natural environments [133,134]. These *Saccharomyces* species can play a major role in food or beverage fermentation. However, the ale yeasts involved in alcoholic fermentation mostly belong to the species *S. cerevisiae*[132]. Besides its important role in baking and brewing, this yeast species has been used as a eukaryotic model organism in molecular and cell biology, for example, the characteristic of many proteins can be discovered by studying their homologs in *S. cerevisiae*.

S. cerevisiae cells are round to ovoid, are 5 to 10 μm in diameter, reproduce by a budding process and can grow aerobically on glucose, maltose and trehalose but not on lactose or cellobiose. In the presence of oxygen, it is even able to operate in a mixed fermentation/respiration mode. The ratio of fermentation to respiration varies slightly among strains but is approximately 80:20 [135]. Furthermore, *S. cerevisiae* can be processed to produce potential advanced biofuels such as long-chain alcohols and isoprenoid- and fatty acid-based biofuels, which have physical properties that more closely resemble petroleum-derived fuels [136].

Advantages of Yeast for MFC

Yeast is sometimes thought to be impractical as a biocatalyst, due to difficulties with transferring electrons out of cellular organelles [137]. However, since yeasts are robust, are easily handled, are mostly non-pathogenic, have high catabolic rates and grow on substrate spectrum, they are well worth considering as promising biocatalysts for MFCs [138]. In addition, several other merits may exist for using *S. cerevisiae* in MFCs. First, *S. cerevisiae* can survive and function in an anaerobic condition that is required for the anode compartment of traditional MFC. Second, the optimal growth temperature for *S. cerevisiae* is around 30°C, which is a convenient ambient temperature. Third, the metabolism of this species is well understood, which helps locate mechanisms responsible for electricity generation in MFCs. Lastly, yeast-based fuel cells could be retrofitted into ethanol plants for *in situ* power generation [139].

Yeast for in Situ Power Generation

In an anaerobic condition, yeasts usually switch to fermentation reactions where one glucose molecule is consumed for the production of two molecules of pyruvates. Pyruvate is further transformed into alcohol or organic acid by recycling NADH to NAD⁺, which is a key step to sustain the glycolysis process [140]. This glycolysis reaction takes place in the cytosol of the cell rather than in the mitochondria, so NADH could be easily accessed by the mediator molecule present in the cell membrane of the yeast [73]. The glycolysis and the oxidation of NADH to NAD⁺ are not influenced by the energy extraction process in the MFCs. Based on these characteristics, MFCs using yeast can be directly applied in fermenters for *in situ* power generation [139].

Limitation of S. Cerevisiae for MFC use

Limitations exist for *S. cerevisiae* to be used in MFCs. First, *S. cerevisiae* has a weak ability to oxidise the substrate to supply the maximum number of electrons available for yeast-based MFC. In the mitochondrial process of *S. cerevisiae*, there is a total of only 14 ATP per glucose molecule produced, which is much less than a net of 28 to 30 ATP typically achieved by most aerobes [138]. Also, mediators are commonly required to facilitate the transfer of electrons to the anode, which makes exogenous mediators necessary to MFCs based on *S. cerevisiae* because this yeast is thought incapable of producing such mediators indigenously [139].

The Output of S. Cerevisiae-based MFCs

In general, yeast-based MFCs perform better than cyanobacteria but still have a lower power output than bacterial fuel cells [138]. It was shown that methylene blue-mediated *S. cerevisiae*MFC can give a power density of 1.5 W m⁻²[141], which is less than the maximum of 6.86 W m⁻²reported by Fan et al. [135] for a mixed-culture MFC. The corresponding volumetric density, based on the specified anodic chamber volume of 10 ml, is 90 W m⁻³.

A recent MFC that employs *S. cerevisiae* as the electron donor in the anodic half-cell and *C. vulgaris* as the electron acceptor in the cathodic half-cell can reach a maximum power at 90 mV and a load of 5,000 Ω, giving a power density of 0.95 mW m^{-2} of electrode surface area [142]. This power density is still very low.

Another study investigated the possibility of *S. cerevisiae* to transfer electrons to an extracellular electron acceptor through DET mode and found that the cells that adhered to the anode were able to sustain power generation in a mediator-less MFC. However, the power performance of this MFC was extremely low (0.003 W m^{-2}) [143].

Geobacter Sulfurreducens

G. sulfurreducens are comma-shaped Gram-negative, anaerobic bacteria capable of coupling oxidization of organic compounds to reduction of metals. This organism is one of the predominant metal-reducing bacteria in soil and hence plays an important ecological role in biotechnologically exploitable bioremediation. The activity of *Geobacter* species in sub-surface can be stimulated to remove organic and metal contaminants such as aromatic hydrocarbons and uranium from groundwater [144-146].

The genome sequence of *G. sulfurreducens* is available, and a system for genetic manipulation has been developed for this organism [147]. Since it was discovered in 1994 [148], this bacterium has been extensively studied for MFC applications. It has been reported that (1) *G. sulfurreducens* can completely oxidise electron donors by using only an electrode as the electron acceptor, (2) it can quantitatively transfer electrons to electrodes in the absence of electron mediators, and (3) this electron transfer is similar to those observed for electron transport to Fe (III) citrate [47].

Advantages of G. Sulfurreducens

G. sulfurreducens is the most abundant species on anode surfaces in MFCs grown with more than one bacterial species [149-151]. It can form biofilms on the anodes, which make all the cells participate in electron transport to the anode and thus increase the current production [152]. *G. sulfurreducens* is an anaerobe but can withstand low levels of oxygen and may use oxygen as an electron acceptor to support growth under aerobic conditions [153].

This *Geobacter* species can produce large amounts of electrical energy since it possesses multiple mechanisms that involve either pili or c-type cytochromes to facilitate the electron transfer to electrode in MFCs (discussed before in the 'Microbial fuel cells' section) [149]. Also, with the electron transfer to electrodes, the *Geobacter* species can effectively oxidise acetate [47,154]. A current density of 4.56 A m^{-2}, corresponding to power densities of 1.88 W m^{-2} and 43 W m^{-3}, measured for *G. sulfurreducens* is among the highest reported for a pure culture [155]. By reducing the anode compartment volume to a fraction of a millilitre, the volumetric density was in fact increased to 2.15 kW m^{-3}. While the lower value is more realistic for comparison to other studies, this does show that very high densities are achievable in principle with this organism. In addition, *G. sulfurreducens* converts acetate to current with coulombic efficiencies of over 90%[151,155].

Previous studies have shown that when a high selective pressure for high rates of current production at high coulombic efficiencies is imposed on complex microbial communities, it is the organisms closely related to *G. sulfurreducens* that are routinely enriched on anodes of the MFCs[55,154,156-158]. Thus, *G. sulfurreducens* can also be used to study adaptation for enhanced power production.

Metabolism of Geobacter Species

The metabolism of *G. sulfurreducens* was investigated by constraint-based modelling [159]. In contrast to *Escherichia coli*, which primarily produces energy and biosynthetic precursors through sugar fermentation, *Geobacter* completely oxidises acetate and other electron donors via the TCA cycle [160,161], which makes it necessary to transfer electrons to terminal electron acceptors for regeneration of cytoplasmic and intramembrane electron acceptors and ATP synthesis. In *G. sulfurreducens*, this is accomplished by electron transfer to extracellular electron acceptors, i.e. Fe(III) oxides [162].

Since the rate of cytoplasmic proton consumption is lower than that of proton production during the reduction of extracellular electron acceptors such as Fe(III), the energy consumption with extracellular electron acceptors is lower compared to that associated with intracellular acceptors[159]. The use of extracellular electron shuttles makes the *Geobacter* species circumvent the metabolic cost of producing the electron shuttles and consequently more energetically competitive than shuttle-producing Fe(III) reducers in sub-surface environments [159].

In silico analysis suggested that the metabolic network of *G. sulfurreducens* contains pyruvate-ferredoxin oxidoreductase, which catalyzes synthesis of pyruvate from acetate and carbon dioxide in a single step, indicating that the synthesis of amino acids in *G. sulfurreducens* is more efficient than in *E. coli*[159].

Limits and Applications

MFCs powered by *G. sulfurreducens* are far away from being commercialized as a practical biofuel source [152], because up until now the current levels of these MFCs are around 14 mA which could be used to power very simple components [149] but still not big enough to drive complex mechanisms. However, the actual current densities that could be generated from MFCs based on*G. sulfurreducens* are still unclear and require further investigation [151].

Comparison of Geobacter sp. and Shewanella sp

Geobacteraceae and Shewanellaceae are classic models in MFC research as their metabolism and versatility have been studied extensively [72,163]. As mentioned before in the 'Microbial fuel cells' section, they are both capable of being used for DET mode in MFCs, because both *Shewanella* sp. and *Geobacter* sp. possess nanowires, electrically conductive bacterial appendages, to transport electrons from cells to solid electron acceptors such as graphite anodes in MFCs [34,35,162]. Despite those similarities, differences also exist when compared regarding the engineering design and performance of the MFCs.

Geobacter-based MFCs generate high coulombic efficiencies [164] but require strict anaerobic conditions which limit their applicability. In contrast, *Shewanella*-based MFCs can be operated with air-exposed cultures [27]. Unlike *Geobacter* sp. that requires direct contact to the electrode surface [72], *Shewanella* sp. can use additional mediators to facilitate electron transfer outside the cell membrane [27]. Importantly, besides utilizing nanowires to mediate the electron transport[165], they can synthesize their own redox mediators (i.e. flavins) for extracellular electron transfer under diverse environmental conditions [163,166]. These two electron-mediated mechanisms determine the efficiency of the current generation in *Shewanella*-containing MFCs[167].

A maximum power density of 24 mW m^{-2} (in the presence of an additional mediator) was reported for *Shewanella*[168]. This value appears low in comparison to bacterial cells, but that is because it was referred to the true microscopic area of a porous electrode. When expressed, as customary, in terms of macroscopic MFC dimensions, the equivalent power densities are 3 W m^{-2} and 500 W m^{-3}. These values compare favourably with *Geobacter*. When dissolved oxygen was deliberately fed into the anode chamber, *Shewanella*-based MFC was still able to produce a power output of 6.5 mW m^{-2} and 13 mA m^{-2}. The MFC used lactate as the fuel source and relied on self-excreted mediators of *Shewanella*[169].

In fact, the previously described *Shewanella*-based system would not be directly applicable to powering electronics and is required to use aerobic water [170]. For instance, a complex pumping system is necessary to continuously recirculate the anolyte between the anode and the large anolyte reservoir, but this pumping system at the anode could consume more power than the*Shewanella*-based MFC produces. Since *Shewanella* sp. cannot use oxygen as the electron acceptor, ferricyanide needs to be added as catholyte. However, ferricyanide is a non-renewable and toxic electron acceptor and can thus not be deployed in the field in the long term. Moreover, the coulombic efficiencies were found to be low (<6%), when calculated based on the incomplete oxidation of lactate to acetate [170].

Conversely, *G. sulfurreducens* can effectively oxidise acetate with electron transfer to electrodes[47,154] and convert acetate to current with coulombic efficiencies of more than 90% [151,155].*G. sulfurreducens* is an anaerobe that can withstand low levels of oxygen and may use oxygen as an electron acceptor [153]. It has recently been shown that with a new configuration, MFCs based on *G. sulfurreducens* can become 100% aerobic, allowing for floating and/or untethered applications. At the same time, the performance of the MFCs is similar to their anaerobic/aerobic counterparts [170]. It is expected that with this aerobic configuration, power could be produced in a *G. sulfurreducens* MFC suspended in aerobic seawater [170].

CONCLUSIONS

Electricity generation in MFCs is based on the metabolic activity of living microorganisms at the anode. The selection of microorganisms is based on many criteria, but the power output, electron transfer ability and biological functions such as photosynthesis and hydrogen production are particularly important. These important properties, in different combinations, are exemplified by the four representative microorganisms discussed above, and their referential facts for modelling are compared in Table 2. Studying these individually, and in combination, should reveal significant insights in the quest for higher power output MFCs.

Most MFC researchers have been active in engineering designs, i.e. how to create scalable and economical architectures and engineer more efficient hardware and how different microbes interact with the anodes/cathodes when transporting electrons [127]. Such research covers optimizing anodic conditions, housing constructs and component materials, learning more about microbial community ecology and isolating vigorous biocatalysts [9]. Biological aspects of MFCs have also received some attention, such as the anodic activity of different organisms. However, very little research has been done on the biochemical interface between the engineering design and the biological aspects (see Figure 1).

We conclude that future studies are required to work on that interface, i.e. how to enhance the anodic activity by means of adjusting the metabolic activity of biocatalysts, for example, utilizing metabolic network analysis. The genome-scale metabolic networks are quite new concepts and have only been produced in the last few years. The analysis of the metabolic network through modelling approaches, such as flux balance analysis, plays an important role in filling the gap between genotypes and phenotypes of microorganisms to provide a full picture of the biological system.

AUTHORS' CONTRIBUTIONS

LM and WV co-conceived the idea. LM conducted the review and drafted the manuscript. WV supervised the work and corrected the manuscript. Both authors read and approved the final manuscript.

REFERENCES

1. Potter, MC: Electrical effects accompanying the decomposition of organic compounds. Proc. R. Soc. London Series B. 84(571), 260–276 (1911).

2. Rittmann, BE: Opportunities for renewable bioenergy using microorganisms. Biotechnol. Bioeng.. 100(2), 203–212 (2008).

3. Pandyaswargo, A, Onoda, H, Nagata, K: Energy recovery potential and life cycle impact assessment of municipal solid waste management technologies in Asian countries using ELP model. Int. J. Energy Environ. Eng.. 3(1), 28 (2012).

4. Eddine, B, Salah, M: Solid waste as renewable source of energy: current and future possibility in Algeria. Int. J. Energy Environ. Eng.. 3(1), 17 (2012)

5. Sharma, V, Kundu, PP: Biocatalysts in microbial fuel cells. Enzyme Microb. Technol.. 47(5), 179–188 (2010).

6. Harnisch, F, Schröder, U: From MFC to MXC: chemical and biological cathodes and their potential for microbial bioelectrochemical systems. Chem. Soc. Rev.. 39(11), 4433–4448 (2010).

7. Girguis, PR, Nielsen, ME, Figueroa, I: Harnessing energy from marine productivity using bioelectrochemical systems. Curr. Opin. Biotechnol.. 21(3), 252–258 (2010).

8. Li, H, Opgenorth, PH, Wernick, DG, Rogers, S, Wu, T-Y, Higashide, W, Malati, P, Huo, Y-X, Cho, KM, Laio, JC: Integrated electromicrobial conversion of CO_2 to higher alcohols. Science. 335(6076), 1596 (2012).

9. Schwartz, K: Microbial fuel cells: design elements and application of a novel renewable energy source. MMG 445 Basic Biotechnology eJournal. 3(1), 20–27 (2007)

10. Logan, BE: Scaling up microbial fuel cells and other bioelectrochemical systems. Appl. Microbiol. Biotechnol.. 85(6), 1665–1671 (2010).

11. Bullen, RA, Arnot, TC, Lakeman, JB, Walsh, FC: Biofuel cells and their development. Biosens. Bioelectron.. 21(11), 2015–2045 (2006).

12. Rupprecht, J: From systems biology to fuel–*Chlamydomonas reinhardtii* as a model for a systems biology approach to improve biohydrogen production. J. Biotechnol.. 142(1), 10–20 (2009).

13. Mukhopadhyay, A, Redding, AM, Rutherford, BJ, Keasling, JD: Importance of systems biology in engineering microbes for biofuel production. Curr. Opin. Biotechnol.. 19(3), 228–234 (2008).

14. Price, ND, Reed, JL, Palsson, BO: Genome-scale models of microbial cells: evaluating the consequences of constraints. Nature Reviews Microbiology. 2(11), 886–897 (2004).

15. Rocha, I, Forster, J, Nielsen, J: Design and application of genome-scale reconstructed metabolic models. Methods Mol. Biol.. 416, 409–431 (2008).

16. Jaqaman, K, Danuser, G: Linking data to models: data regression. Nature Reviews Molecular Cell Biology. 7(11), 813–819 (2006).

17. Terzer, M, Maynard, ND, Covert, MW, Stelling, J: Genome-scale metabolic networks. Wiley Interdiscip. Rev. Syst. Biol. Med.. 1(3), 285–297 (2009).

18. Lee, JM, Gianchandani, EP, Eddy, JA, Papin, JA: Dynamic analysis of integrated signaling, metabolic, and regulatory networks. PLoS Comput. Biol.. 4(5), e1000086 (2008).

19. Covert, MW, Xiao, N, Chen, TJ, Karr, JR: Integrating metabolic, transcriptional regulatory and signal transduction models in *Escherichia coli*. Bioinformatics. 24(18), 2044–2050 (2008).

20. Jamshidi, N, Palsson, B: Mass action stoichiometric simulation models: incorporating kinetics and regulation into stoichiometric models. Biophys. J.. 98(2), 175–185 (2010).

21. Thiele, I, Palsson, BO: A protocol for generating a high-quality genome-scale metabolic reconstruction. Nat. Protoc.. 5(1), 93–121 (2010).

22. Orth, JD, Thiele, I, Palsson, BO: What is flux balance analysis?. Nat Biotech. 28(3), 245–248 (2010).

23. Fernie, AR, Geigenberger, P, Stitt, M: Flux an important, but neglected, component of functional genomics. Curr. Opin. Plant Biol.. 8(2), 174–182 (2005).

24. Saha, R, Suresha, S, Park, W, Lee, D-Y, Karimi, IA: Strain improvement and mediator selection for microbial fuel cell by genome scale in silico model. In: Pleşu V, Agachi PS (eds.) 17th

European Symposium on Computer Aided Process Engineering – ESCAPE17, Bucharest (2007)

25. Rabaey, K, Verstraete, W: Microbial fuel cells: novel biotechnology for energy generation. Trends Biotechnol.. 23(6), 291–298 (2005).

26. Du, Z, Li, H, Gu, T: A state of the art review on microbial fuel cells: a promising technology for wastewater treatment and bioenergy. Biotechnol. Adv.. 25(5), 464–482 (2007).

27. Schröder, U: Anodic electron transfer mechanisms in microbial fuel cells and their energy efficiency. Phys. Chem. Chem. Phys.. 9(21), 2619–2629 (2007).

28. Kim, HJ, Hyun, MS, Chang, IS, Kim, BH: A microbial fuel cell type lactate biosensor using a metal-reducing bacterium, *Shewanella putrefaciens*. J. Microbiol. Biotechnol.. 9(3), 365–367 (1999)

29. Kim, HJ, Park, HS, Hyun, MS, Chang, IS, Kim, M, Kim, BH: A mediator-less microbial fuel cell using a metal reducing bacterium, *Shewanella putrefaciens*. Enzyme Microb. Technol..30(2), 145–152 (2002).

30. Lovley, DR: Microbial fuel cells: novel microbial physiologies and engineering approaches. Curr. Opin. Biotechnol.. 17(3), 327–332 (2006).

31. Chaudhuri, SK: Lovley. DR: Electricity generation by direct oxidation of glucose in mediatorless microbial fuel cells. Nat. Biotechnol.. 21(10), 1229–1232 (2003)

32. Chang, IS, Moon, H, Bretschger, O, Jang, JK, Park, HI, Nealson, KH, Kim, BH: Electrochemically active bacteria (EAB) and mediator-less microbial fuel cells. J. Microbiol. Biotechnol.. 16(2), 163–177 (2006)

33. Schaetzle, O, Barriere, F, Baronian, K: Bacteria and yeasts as catalysts in microbial fuel cells: electron transfer from micro-organisms to electrodes for green electricity. Energy Environ. Sci.. 1(6), 607–620 (2008).

34. Gorby, YA, Yanina, S, McLean, JS, Rosso, KM, Moyles, D, Dohnalkova, A, Beveridge, TJ, Chang, IS, Kim, BH, Kim, KS, Culley, DE, Reed, SB, Romine, MF, Saffarini, DA, Hill, EA, Shi, L, Elias, DA, Kennedy, DW, Pinchuk, G, Watanabe, K, Ishii, S, Logan, B, Nealson, KH, Fredrickson, JK: Electrically conductive bacterial nanowires produced by *Shewanella oneidensis* strain

MR-1 and other microorganisms. Proc. Natl. Acad. Sci.. 103(30), 11358–11363 (2006).

35. Reguera, G, Nevin, KP, Nicoll, JS, Covalla, SF, Woodard, TL, Lovely, DR: Biofilm and nanowire production leads to increased current in *Geobacter sulfurreducens* fuel cells. Appl. Environ. Microbiol.. 72(11), 7345–7348 (2006).

36. Li, P, Ki, J-P, Liu, H: Analysis and optimization of current collecting systems in PEM fuel cells. Int. J. Energy Environ. Eng.. 3(1), 2 (2012).

37. Cooney, MJ, Roschi, E, Marison, IW, Comninellis, C, von Stockar, U: Physiologic studies with the sulfate-reducing bacterium *Desulfovibrio desulfuricans*: evaluation for use in a biofuel cell. Enzyme Microb. Technol.. 18(5), 358–365 (1996).

38. Scholz, F, Schröder, U: Bacterial batteries. Nat. Biotechnol.. 21(9), 3–4 (2003).

39. Rabaey, K, Lissens, G, Verstraete, W: Microbial fuel cells: performances and perspectives. In: Lens P, Westermann P, Haberbauer M, Moreno A (eds.) Biofuels for Fuel Cells: Biomass Fermentation Towards Usage in Fuel Cells, pp. 377–399. London: IWA (2005)

40. Rosenbaum, M, He, Z, Angenent, LT: Light energy to bioelectricity: photosynthetic microbial fuel cells. Curr. Opin. Biotechnol.. 21(3), 259–264 (2010).

41. Davis, F, Higson, SPJ: Biofuel cells–recent advances and applications. Biosens. Bioelectron..22(7), 1224–1235 (2007).

42. Bennetto, HP, Delany, GM, Mason, JR, Roller, SD, Stirling, JL, Thurston, CF: The sucrose fuel cell: efficient biomass conversion using a microbial catalyst. Biotechnol. Lett.. 7(10), 699–704 (1985).

43. Rabaey, K, Boon, N, Höfte, M, Verstraete, W: Microbial phenazine production enhances electron transfer in biofuel cells. Environ. Sci. Technol.. 39(9), 3401–3408 (2005).

44. Park, HS, Kim, BH, Kim, HS, Kim, HJ, Kim, GT, Kim, M, Chang, IS, Park, YK, Chang, HI: A novel electrochemically active and Fe(III)-reducing bacterium phylogenetically related to*Clostridium butyricum* isolated from a microbial fuel cell. Anaerobe. 7(6), 297–306 (2001).

45. Kim, BH, Park, HS, Kim, HJ, Kim, GT, Chang, IS, Lee, J, Phung, NT: Direct electrode reaction of Fe(III)-reducing bacterium, *Shewanella putrefaciens*. J. Microbiol. Biotechnol..9(2), 127–131 (1999)

46. Logan, BE, Regan, JM: Electricity-producing bacterial communities in microbial fuel cells. Trends Microbiol.. 14(12), 512–518 (2006).

47. Bond, DR, Lovley, DR: Electricity production by Geobacter sulfurreducens attached to electrodes.. Appl. Environ. Microbiol.. 69(3), 1548–1555 (2003).

48. Lanthier, M, Gregory, KB, Lovley, DR: Growth with high planktonic biomass in *Shewanella oneidensis* fuel cells. FEMS Microbiol. Lett.. 278(1), 29–35 (2008).

49. Logan, BE, Regan, JM: Microbial fuel cells, challenges and applications. Environ. Sci. Technol.. 40(17), 5172–5180 (2006).

50. Fan, Y, Hu, H, Liu, H: Enhanced Coulombic efficiency and power density of air-cathode microbial fuel cells with an improved cell configuration. J. Power. Sources. 171(2), 348–354 (2007).

51. Zhang, F, Cheng, S, Pant, D, Van Bogaert, G, Logan, BE: Power generation using an activated carbon and metal mesh cathode in a microbial fuel cell. Electrochem. Commun..11(11), 2177–2179 (2009).

52. Zuo, Y, Cheng, S, Logan, BE: Ion exchange membrane cathodes for scalable microbial fuel cells. Environ. Sci. Technol.. 42(18), 6967–6972 (2008).

53. Chen, G-W, Choi, SJ, Lee, TH, Lee, GY, Cha, JH, Kim, CW: Application of biocathode in microbial fuel cells: cell performance and microbial community. Appl. Microbiol. Biotechnol..79(3), 379–388 (2008).

54. Cheng, S, Logan, BE: Ammonia treatment of carbon cloth anodes to enhance power generation of microbial fuel cells. Electrochem. Commun.. 9, 492–496 (2006)

55. Liu, JL, Lowy, DA, Baumann, RG, Tender, LM: Influence of anode pretreatment on its microbial colonization. J. Appl. Microbiol.. 102(1), 177–183 (2007).

56. Cao, X, Huang, X, Boon, N, Liang, P, Fan, M: Electricity generation by an enriched phototrophic consortium in a microbial fuel cell. Electrochem. Commun.. 10(9), 1392–1395 (2008).

57. Yang, Y, Sun, G, Xu, M: Microbial fuel cells come of age. J. Chem. Technol. Biotechnol..86(5), 625–632 (2011).

58. Lovley, DR: Extracellular electron transfer: wires, capacitors, iron lungs, and more. Geobiology. 6(3), 225–231 (2008).

59. Shi, L, Squier, TC, Zachara, JM, Fredrickson, JK: Respiration of metal (hydr)oxides by Shewanella and Geobacter: a key role for multihaem c-type cytochromes. Mol. Microbiol..65(1), 12–20 (2007).

60. Oh, ST, Kim, JR, Premier, GC, Lee, TH, Kim, C, Sloan, WT: Sustainable wastewater treatment: how might microbial fuel cells contribute. Biotechnol. Adv.. 28(6), 871–881 (2010).

61. Yuan, Y, Ahmed, J, Zhou, L, Zhao, B, Kim, S: Carbon nanoparticles-assisted mediator-less microbial fuel cells using Proteus vulgaris. Biosens. Bioelectron.. 27(1), 106–112 (2011).

62. He, Z, Kan, J, Mansfeld, F, Angenent, LT, Nealson, KH: Self-sustained phototrophic microbial fuel cells based on the synergistic cooperation between photosynthetic microorganisms and heterotrophic bacteria. Environ. Sci. Technol.. 43(5), 1648–1654 (2009).

63. Malik, S, Drott, E, Grisdela, P, Lee, J, Lee, C, Lowy, DA, Gray, S, Tender, LM: A self-assembling self-repairing microbial photoelectrochemical solar cell. Energy Environ. Sci..2(3), 292–298 (2009).

64. Cho, YK, Donohue, TJ, Tejedor, I, Anderson, MA, McMahon, KD, Noguera, DR: Development of a solar-powered microbial fuel cell. J. Appl. Microbiol.. 104(3), 640–650 (2008).

65. Rosenbaum, M, Schröder, U, Scholz, F: In situ electrooxidation of photobiological hydrogen in a photobioelectrochemical fuel cell based on Rhodobacter sphaeroides. Environ. Sci. Technol.. 39(16), 6328–6333 (2005).

66. Xing, D, Zuo, Y, Cheng, S, Regan, JM, Logan, BE: Electricity generation by Rhodopseudomonas palustris DX-1. Environ. Sci. Technol.. 42(11), 4146–4151 (2008).

67. Larsen, K, Ibrom, A, Beier, C, Jonasson, S, Michelsen, A: Ecosystem respiration depends strongly on photosynthesis in a temperate heath. Biogeochemistry. 85(2), 201–213 (2007).

68. Stal, LJ, van Gemerden, H, Krumbein, WE: Structure and development of a benthic marine microbial mat. FEMS Microbiol. Lett.. 31(2), 111–125 (1985).

69. Strik, D, Terlouw, H, Hamelers, HV, Buisman, CJ: Renewable sustainable biocatalyzed electricity production in a photosynthetic algal microbial fuel cell (PAMFC). Appl. Microbiol. Biotechnol.. 81(4), 659–668 (2008).

70. Zou, Y, Pisciotta, J, Billmyre, RB, Baskakov, IV: Photosynthetic microbial fuel cells with positive light response. Biotechnol. Bioeng.. 104(5), 939–946 (2009).

71. Yagishita, T, Horigome, T, Tanaka, K: Effects of light, CO_2 and inhibitors on the current output of biofuel cells containing the photosynthetic organism *Synechococcus* sp. J. Chem. Technol. Biotechnol.. 56(4), 393–399 (1993)

72. Lovley, DR: Bug juice: harvesting electricity with microorganisms. Nature Reviews. Microbiology. 4(7), 497–508 (2006).

73. Raghavulu, SV, Goud, RK, Sarma, PN, Mohan, SV: *Saccharomyces cerevisiae* as anodic biocatalyst for power generation in biofuel cell: influence of redox condition and substrate load. Bioresour. Technol.. 102(3), 2751–2757 (2011).

74. Rosenbaum, M, Schröder, U: Photomicrobial solar and fuel cells. Electroanalysis. 22(7–8), 844–855 (2010)

75. Dal'Molin, CG, Quek, LE, Palfreyman, RW, Nielsen, LK: AlgaGEM - a genome-scale metabolic reconstruction of algae based on the *Chlamydomonas reinhardtii* genome. BMC Genomics.12(Suppl 4), S5 (2011).

76. Chang, RL, Ghamsari, L, Manichaikul, A, Hom, EF, Balaji, S, Fu, W, Shen, Y, Hao, T, Palsson, BØ, Salehi-Ashtiani, K, Papin, JA: Metabolic network reconstruction of *Chlamydomonas* offers insight into light-driven algal metabolism. Mol. Syst. Biol.. 7, 518 (2011).

77. Yoshikawa, K, Kojima, Y, Nakajima, T, Furusawa, C, Hirasawa, T, Shimizu, H: Reconstruction and verification of a genome-scale metabolic model for *Synechocystis* sp. PCC6803. Appl. Microbiol. Biotechnol.. 92(2), 347–358 (2011).

78. Nogales, J, Gudmundsson, S, Knight, EM, Palsson, BO, Thiele, I: Detailing the optimality of photosynthesis in cyanobacteria through systems biology analysis. Proc. Natl. Acad. Sci..109(7), 2678–2683 (2012).

79. Sun, J, Sayyar, B, Butler, JE, Pharkya, P, Fahland, TR, Famili, I, Schilling, CH, Lovley, DR, Mahadevan, R: Genome-scale constraint-based modeling of *Geobacter metallireducens*. BMC Syst. Biol.. 3, 15 (2009).

80. Heavner, BD, Smallbone, K, Barker, B, Mendes, P, Walker, LP: Yeast 5 - an expanded reconstruction of the *Saccharomyces cerevisiae* metabolic network. BMC Syst. Biol.. 6, 55 (2012).

81. Adams, M: General growth and mating.

82. Sheng, J, Kim, HW, Badalamenti, JP, Zhou, C, Sridharakrishnan, S, Krajmalnik-Brown, R, Rittmann, BE, Vannela, R: Effects of temperature shifts on growth rate and lipid characteristics of *Synechocystis* sp. PCC6803 in a bench-top photobioreactor. Bioresour. Technol.. 102(24), 11218–11225 (2011).

83. Watson, K: Temperature relations. In: Rose AH, Harrison JS (eds.) The Yeasts, pp. 41–71. London: Academic (1987)

84. Trinh, N, Park, J, Kim, B-W: Increased generation of electricity in a microbial fuel cell using*Geobacter sulfurreducens*. Korean J. Chem. Eng.. 26(3), 748–753 (2009).

85. Yoon, HS, Hackett, JD, Ciniglia, C, Pinto, G, Bhattacharya, D: A molecular timeline for the origin of photosynthetic eukaryotes. Mol. Biol. Evol.. 21(5), 809–818 (2004).

86. Merchant, SS, Prochnik, SE, Vallon, O, Harris, EH, Karpowicz, SJ, Witman, GB, Terry, A, Salamov, A, Fritz-Laylin, LK, Maréchal-Drouard, L, Marshall, WF, Qu, LH, Nelson, DR, Sanderfoot, AA, Spalding, MH, Kapitonov, VV, Ren, Q, Ferris, P, Lindquist, E, Shapiro, H, Lucas, SM, Grimwood, J, Schmutz, J, Cardol, P, Cerutti, H, Chanfreau, G, Chen, CL, Cognat, V, Croft, MT, Dent, R: The *Chlamydomonas* genome reveals the evolution of key animal and plant functions. Science. 318(5848), 245–250 (2007).

87. Harris, EH: *Chlamydomonas* as a model organism. Annu. Rev. Plant Physiol. Plant Mol. Biol..52(1), 363–406 (2001).

88. Manichaikul, A, Ghamsari, L, Hom, EF, Lin, C, Murray, RR, Chang, RL, Balaji, S, Hao, T, Shen, Y, Chavali, AK, Thiele, I, Yang, X, Fan, C, Mello, E, Hill, DE, Vidal, M, Salehi-Ashtiani, K, Papin, JA: Metabolic network analysis integrated with transcript verification for sequenced genomes. Nat. Methods. 6(8), 589–592 (2009).

89. Keller, LC, Romijn, EP, Zamora, I, Yates, JR 3rd., Marshall, WF: Proteomic analysis of isolated chlamydomonas centrioles reveals orthologs of ciliary-disease genes. Current Biol.: CB. 15(12), 1090–1098 (2005).

90. Pazour, GJ, Agrin, N, Walker, BL, Witman, GB: Identification of predicted human outer dynein arm genes: candidates for primary ciliary dyskinesia genes. J. Med. Genet.. 43(1), 62–73 (2006).

91. Vilchez, C, Garbayo, I, Markvicheva, E, Galván, F, León, R: Studies on the suitability of alginate-entrapped *Chlamydomonas reinhardtii* cells for sustaining nitrate consumption processes. Bioresour. Technol.. 78(1), 55–61 (2001).

92. Ghirardi, ML, Posewitz, MC, Maness, PC, Dubini, A, Yu, J, Seibert, M: Hydrogenases and hydrogen photoproduction in oxygenic photosynthetic organisms. Annu. Rev. Plant Biol.. 58, 71–91 (2007).

93. Kosourov, SN, Seibert, M: Hydrogen photoproduction by nutrient-deprived Chlamydomonas reinhardtii cells immobilized within thin alginate films under aerobic and anaerobic conditions.. Biotechnol. Bioeng.. 102(1), 50–58 (2009).

94. The Chlamydomonas Center: Chlamydomonas connection

95. Chen, P, Min, M, Chen, Y, Wang, L, Li, Y, Chen, Q, Wang, C, Wan, Y, Wang, X, Cheng, Y, Deng, S, Hennessy, K, Lin, X, Liu, Y, Wang, Y, Martinez, B, Ruan, R: Review of biological and engineering aspects of algae to fuels approach. Int. J. Agric. Biol. Eng.. 2(4), 1–30 (2009)

96. Pienkos, PT, Darzins, A: The promise and challenges of microalgal-derived biofuels. Biofpr..3, 431–440 (2009)

97. Campbell, MN: Biodiesel: algae as a renewable source for liquid fuel. Guelph Eng. J..1(1916–1107), 207 (2008)

98. Velasquez-Orta, SB, Curtis, TP, Logan, BE: Energy from algae using microbial fuel cells. Biotechnol. Bioeng.. 103(6), 1068–1076 (2009).

99. Rosenbaum, M, Schröder, U, Scholz, F: Utilizing the green alga *Chlamydomonas reinhardtii* for microbial electricity generation: a living solar cell. Appl. Microbiol. Biotechnol.. 68(6), 753–756 (2005).

100. Strik, DPBTB, Hamelers, HVM, Buisman, CJN: Solar energy powered microbial fuel cell with a reversible bioelectrode. Environ. Sci. Technol.. 44(1), 532–537 (2009)

101. Harnisch, F, Schröder, U: Selectivity versus mobility: separation of anode and cathode in microbial bioelectrochemical systems. ChemSusChem. 2(10), 921–926 (2009).

102. Melis, A, Happe, T: Hydrogen production. Green algae as a source of energy. Plant Physiol..127(3), 740–748 (2001).

103. Rupprecht, J, Hankamer, B, Mussgnug, JH, Ananyev, G, Dismukes, C, Kruse, O: Perspectives and advances of biological H_2 production in microorganisms. Appl. Microbiol. Biotechnol.. 72(3), 442–449 (2006).

104. Hankamer, B, Lehr, F, Rupprecht, J, Mussgnug, JH, Posten, C, Kruse, O: Photosynthetic biomass and H2 production by green algae: from bioengineering to bioreactor scale-up. Physiol. Plant.. 131(1), 10–21 (2007).

105. Gaffron, H, Rubin, J: Fermentative and photochemical production of hydrogen in algae. J. Gen. Physiol.. 26(2), 219–240 (1942).

106. Kruse, O, Rupprecht, J, Mussgnug, JH, Dismukes, GC, Hankamer, B: Photosynthesis: a blueprint for solar energy capture and biohydrogen production technologies. Photochem. Photobiol. Sci.. 4(12), 957–970 (2005).

107. Melis, A, Zhang, L, Forestier, M, Ghirardi, ML, Seibert, M: Sustained photobiological hydrogen gas production upon reversible inactivation of oxygen evolution in the green alga*Chlamydomonas reinhardtii*. Plant Physiol.. 122(1), 127–136 (2000).

108. Ghirardi, ML, Zhang, L, Lee, JW, Flynn, T, Seibert, M, Greenbaum, E, Melis, A: Microalgae: a green source of renewable H2. Trends Biotechnol.. 18(12), 506–511 (2000).

109. Esqulvel, MG, Amaro, HM, Pinto, TS, Fevereiro, PS, Malcata, FX: Efficient H2 production via*Chlamydomonas reinhardtii.* Trends Biotechnol.. 29(12), 595–600 (2011).

110. Mus, F, Dubini, A, Seibert, M, Posewitz, MC, Grossman, AR: Anaerobic acclimation in*Chlamydomonas reinhardtii.* J. Biol. Chem.. 282(35), 25475–25486 (2007).

111. Dubini, A, Mus, F, Seibert, M, Grossmman, AR, Posewitz, MC: Flexibility in anaerobic metabolism as revealed in a, mutant of Chlamydomonas reinhardtii lacking hydrogenase activity. J. Biol. Chem.. 284(11), 7201–7213 (2009).

112. Posewitz, MC, Dubini, A, Meuser, JE, Seibert, M, Ghirardi, ML: Hydrogenases, hydrogen production, and anoxia. In: Elizabeth HH, Stern DB, Witman GB (eds.) The Chlamydomonas Sourcebook, pp. 217–255. London: Academic (2009)

113. Timmins, M, Thomas-Hall, SR, Darling, A, Zhang, E, Hankamer, B, Marx, UC, Schenk, PM: Phylogenetic and molecular analysis of hydrogen-producing green algae. J. Exp. Bot..60(6), 1691–1702 (2009).

114. Doebbe, A, Keck, M, La Russa, M, Mussgnug, JH, Hankamer, B, Tekçe, E, Niehaus, K, Kruse, O: The interplay of proton, electron, and metabolite supply for photosynthetic H2 production in *Chlamydomonas reinhardtii.* J. Biol. Chem.. 285(39), 30247–30260 (2010).

115. Grossman, AR, Catalanotti, C, Yang, W, Dubini, A, Magneschi, L, Subramanian, V, Posewitz, MC, Seibert, M: Multiple facets of anoxic metabolism and hydrogen production in the unicellular green alga *Chlamydomonas reinhardtii.* New Phytol.. 190(2), 279–288 (2011).

116. Markov, SA, Eivazova, ER, Greenwood, J: Photostimulation of H2 production in the green alga *Chlamydomonas reinhardtii* upon photoinhibition of its O2-evolving system. Int. J. Hydrogen Energy. 31(10), 1314–1317 (2006).

117. Seibert, M, King, PW, Posewitz, MC, Melis, A, Ghirardi, ML: Photosynthetic water-splitting for hydrogen production. In: Wall JD, Harwood CS, Demain A (eds.) Bioenergy, pp. 273–291. Washington D.C: ASM (2008)

118. Zorina, A, Mironov, K, Stepanchenko, N, Sinetova, M, Koroban, N, Zinchenko, V, Kupriyanova, E, Allakhverdiev, S, Los, D: Regulation systems for stress responses in cyanobacteria. Russian J. Plant Physiol. 58(5), 749–767 (2011).

119. Waterbury, JB, Watson, SW, Guillard, RL, Brand, LE: Widespread occurrence of a unicellular, marine, planktonic, cyanobacterium. Nature. 277(5694), 293–294 (1979).

120. Carr, NG, Whitton, BA: The Biology of Cyanobacteria, Berkeley: University of California Press (1982)

121. Martin, W, Rujan, T, Richly, E, Hansen, A, Cornelsen, S, Lins, T, Leister, D, Stoebe, B, Hasegawa, M, Penny, D: Evolutionary analysis of *Arabidopsis*, cyanobacterial, and chloroplast genomes reveals plastid phylogeny and thousands of cyanobacterial genes in the nucleus. Proc. Natl. Acad. Sci. U.S.A.. 99(19), 12246–12251 (2002).

122. Dutta, D, De, D, Chaudhuri, S, Bhattacharya, SK: Hydrogen production by Cyanobacteria. Microb. Cell Fact.. 4, 36 (2005).

123. Kaneko, T, Sato, S, Kotani, H, Tanaka, A, Asamizu, E, Nakamura, Y, Miyajima, N, Hirosawa, M, Sugiura, M, Sasamoto, S, Kimura, T, Hosouchi, T, Matsuno, A, Muraki, A, Nakazaki, N, Naruo, K, Okumura, S, Shimpo, S, Takeuchi, C, Wada, T, Watanabe, A, Yamada, M, Yasuda, M, Tabata, S: Sequence analysis of the genome of the unicellular cyanobacterium*Synechocystis* sp. strain PCC6803. II. Sequence determination of the entire genome and assignment of potential protein-coding regions. DNA Res.. 3(3), 109–136 (1996).

124. Nakamura, Y, Kaneko, T, Hirosawa, M, Miyajima, N, Tabata, S: CyanoBase, a www database containing the complete nucleotide sequence of the genome of *Synechocystis* sp. strain PCC6803. Nucleic Acids Res.. 26(1), 63–67 (1998).

125. Atsumi, S, Higashide, W, Liao, JC: Direct photosynthetic recycling of carbon dioxide to isobutyraldehyde. Nat. Biotechnol.. 27(12), 1177–1180 (2009).

126. Johnson, CH, Stewart, PL, Egli, M: The cyanobacterial circadian system: from biophysics to bioevolution. Annu. Rev. Biophys.. 40, 143–167 (2011).

127. Lovley, DR: The microbe electric: conversion of organic matter to electricity. Curr. Opin. Biotechnol.. 19(6), 564–571 (2008).

128. Pisciotta, JM, Zou, Y, Baskakov, IV: Role of the photosynthetic electron transfer chain in electrogenic activity of cyanobacteria. Appl. Microbiol. Biotechnol.. 91(2), 377–385 (2011).

129. Pisciotta, JM, Zou, Y, Baskakov, IV: Light-dependent electrogenic activity of cyanobacteria. PLoS One. 5(5), e10821 (2010).

130. Madiraju, KS, Lyew, D, Kok, R, Raghavan, V: Carbon neutral electricity production by*Synechocystis* sp. PCC6803 in a microbial fuel cell. Bioresour. Technol.. 110, 214–218 (2012).

131. McCormick, AJ, Bombelli, P, Scott, AM, Philips, AJ, Smith, AG, Fisher, AC, Howe, CJ: Photosynthetic biofilms in pure culture harness solar energy in a mediatorless bio-photovoltaic cell (BPV) system. Energy Environ. Sci.. 4(11), 4699–4709 (2011).

132. Dequin, S, Casaregola, S: The genomes of fermentative *Saccharomyces*. Comptes Rendus Biologies. 334(8–9), 687–693 (2011).

133. Kurtzman, CP: Phylogenetic circumscription of *Saccharomyces*, *Kluyveromyces* and other members of the Saccharomycetaceae, and the proposal of the new genera *Lachancea,Nakaseomyces, Naumovia*. Vanderwaltozyma and Zygotorulaspora. FEMS Yeast Res.. 4(3), 233–245 (2003).

134. Vaughan-Martini, A, Martini, A: Facts, myths and legends on the prime industrial microorganism. J. Ind. Microbiol.. 14(6), 514–522 (1995).

135. Fan, Y, Sharbrough, E, Liu, H: Quantification of the internal resistance distribution of microbial fuel cells. Environ. Sci. Technol. 42(21), 8101–8107 (2008).

136. Peralta-Yahya, PP, Keasling, JD: Advanced biofuel production in microbes. Biotechnol. J..5(2), 147–162 (2010).

137. Wilkinson, S: "Gastrobots" - benefits and challenges of microbial fuel cells in food powered robot applications. Auton. Robots. 9(2), 99–111 (2000).

138. Haslett, ND, Rawson, FJ, Barrière, F, Kunze, G, Pasco, N, Gooneratne, R, Baronian, KH, Haslett, ND, Rawson, FJ, Barrière, F, Kunze, G, Pasco, N, Gooneratne, R, Baronian, KH:

Characterisation of yeast microbial fuel cell with the yeast *Arxula adeninivorans* as the biocatalyst. Biosens. Bioelectron.. 26(9), 3742–3747 (2011).

139. Gunawardena, A, Fernando, S, To, F: Performance of a yeast-mediated biological fuel cell. Int. J. Mol. Sci.. 9(10), 1893–1907 (2008).

140. Feldmann, H: Yeast Molecular Biology: A Short Compendium on Basic Features and Novel Aspects, Munich: Adolf-Butenandt-Institut, University of Munich (2005)

141. Ganguli, R, Dunn, BS: Kinetics of anode reactions for a yeast-catalysed microbial fuel cell. Fuel Cells. 9(1), 44–52 (2009).

142. Powell, EE, Evitts, RW, Hill, GA, Bolster, JC: A microbial fuel cell with a photosynthetic microalgae cathodic half cell coupled to a yeast anodic half cell. Energy Sources Part a-Recovery Utilization Environ. Effects. 33(5), 440–448 (2011).

143. Sayed, ET, Tsujiguchi, T, Nakagawa, N: Catalytic activity of baker's yeast in a mediatorless microbial fuel cell. Bioelectrochemistry. 86, 97–101 (2012).

144. Anderson, RT, Vrionis, HA, Ortiz-Bernad, I, Resch, CT, Long, PE, Dayvault, R, Karp, K, Marutzky, S, Metzler, DR, Peacock, A, White, DC, Lowe, M, Lovley, DR: Stimulating the in situ activity of *Geobacter* species to remove uranium from the groundwater of a uranium-contaminated aquifer. Appl. Environ. Microbiol.. 69(10), 5884–5891 (2003).

145. Holmes, DE, Finneran, KT, O'Neil, RA, Lovley, DR: Enrichment of members of the family Geobacteraceae associated with stimulation of dissimilatory metal reduction in uranium-contaminated aquifer sediments. Appl. Environ. Microbiol.. 68(5), 2300–2306 (2002).

146. Lovley, DR, Anderson, RT: Influence of dissimilatory metal reduction on fate of organic and metal contaminants in the subsurface. Hydrogeol. J.. 8, 77–88 (2000).

147. Coppi, MV, Leang, C, Sandler, SJ, Lovley, DR: Development of a genetic system for *Geobacter sulfurreducens*. Appl. Environ. Microbiol.. 67(7), 3180–3187 (2001).

148. Caccavo, F Jr., Lonergan, DJ, Lovley, DR, Stolz, JF, McInerney, MJ: *Geobacter sulfurreducens* sp. nov., a hydrogen- and acetate-

oxidizing dissimilatory metal-reducing microorganism. Appl. Environ. Microbiol.. 60(10), 3752–3759 (1994).

149. Nevin, KP, Kim, BC, Glaven, RH, Johnson, JP, Woodard, TL, Methé, BA, Didonato, RJ, Covalla, SF, Franks, AE, Liu, A, Lovley, DR: Anode biofilm transcriptomics reveals outer surface components essential for high density current production in *Geobacter sulfurreducens* fuel cells. PLoS One. 4(5), e5628 (2009).

150. Chae, KJ, Choi, MJ, Lee, JW, Kim, KY, Kim, IS: Effect of different substrates on the performance, bacterial diversity, and bacterial viability in microbial fuel cells. Bioresour. Technol.. 100(14), 3518–3525 (2009).

151. Yi, H, Nevin, KP, Kim, BC, Franks, AE, Klimes, A, Tender, LM, Lovley, DR: Selection of a variant of *Geobacter sulfurreducens* with enhanced capacity for current production in microbial fuel cells. Biosens. Bioelectron.. 24(12), 3498–3503 (2009).

152. Salgado, CA: Microbial fuel cells powered by *Geobacter sulfurreducens*. MMG 445 Basic Biotechnology eJournal. 5, 96–101 (2009)

153. Lin, WC, Coppi, MV, Lovley, DR: *Geobacter sulfurreducens* can grow with oxygen as a terminal electron acceptor. Appl. Environ. Microbiol.. 70(4), 2525–2528 (2004).

154. Bond, DR, Holmes, DE, Tender, LM, Lovley, DR: Electrode-reducing microorganisms that harvest energy from marine sediments. Science. 295(5554), 483–485 (2002).

155. Nevin, KP, Richter, H, Covalla, SF, Johnson, JP, Woodard, TL, Orloff, AL, Jia, H, Zhang, M, Lovley, DR: Power output and columbic efficiencies from biofilms of *Geobacter sulfurreducens* comparable to mixed community microbial fuel cells. Environ. Microbiol..10(10), 2505–2514 (2008).

156. Gregory, KB, Bond, DR, Lovley, DR: Graphite electrodes as electron donors for anaerobic respiration. Environ. Microbiol.. 6(6), 596–604 (2004).

157. Holmes, DE, Bond, DR, O'Neil, RA, Reimers, CE, Tender, LR, Lovley, DR: Microbial communities associated with electrodes harvesting electricity from a variety of aquatic sediments. Microb. Ecol.. 48(2), 178–190 (2004).

158. Tender, LM, Reimers, CE, Stecher, HA, Holmes, DE, Bond, DR, Lowy, DA, Pilobello, K, Fertig, SJ, Lovley, DR: Harnessing microbially generated power on the seafloor. Nat. Biotechnol.. 20(8), 821–825 (2002).

159. Mahadevan, R, Bond, DR, Butler, JE, Esteve-Nuñez, A, Coppi, MV, Palsson, BO, Schilling, CH, Lovley, DR: Characterization of metabolism in the Fe(III)-reducing organism *Geobacter sulfurreducens* by constraint-based modeling. Appl. Environ. Microbiol.. 72(2), 1558–1568 (2006).

160. Champine, JE, Underhill, B, Johnston, JM, Lilly, WW, Goodwin, S: Electron transfer in the dissimilatory iron-reducing bacterium *Geobacter metallireducens*. Anaerobe. 6(3), 187–196 (2000).

161. Galushko, AS, Schink, B: Oxidation of acetate through reactions of the citric acid cycle by Geobacter sulfurreducens in pure culture and in syntrophic coculture.. Arch. Microbiol..174(5), 314–321 (2000).

162. Reguera, G, McCarthy, KD, Mehta, T, Nicoll, JS, Tuominen, MT, Lovley, DR: Extracellular electron transfer via microbial nanowires. Nature. 435(7045), 1098–1101 (2005).

163. Fredrickson, JK, Romine, MF, Beliaev, AS, Auchtung, JM, Driscoll, ME, Gardner, TS, Nealson, KH, Osterman, AL, Pinchuk, G, Reed, JL, Rodionov, DA, Rodrigues, JLM, Saffarini, DA, Serres, MH, Spormann, AF, Zhulin, IB, Tiedje, JM: Towards environmental systems biology of *Shewanella*. Nature Reviews Microbiology. 6(8), 592–603 (2008).

164. Call, DF, Wagner, RC, Logan, BE: Hydrogen production by *Geobacter* species and a mixed consortium in a microbial electrolysis cell. Appl. Environ. Microbiol.. 75(24), 7579–7587 (2009).

165. El-Naggar, MY, Gorby, YA, Xia, W, Nealson, KH: The molecular density of states in bacterial nanowires. Biophys. J.. 95(1), L10–L12 (2008).

166. Marsili, E, Baron, DB, Shikhare, ID, Coursolle, D, Gralnick, JA, Bond, DR: *Shewanella*secretes flavins that mediate extracellular electron transfer. Proc. Natl. Acad. Sci.. 105(10), 3968–3973 (2008).

167. Newton, GJ, Mori, S, Nakamura, R, Hashimoto, K, Watanabe, K: Analyses of current-generating mechanisms of *Shewanella loihica* PV-4 and *Shewanella oneidensis* MR-1 in microbial fuel cells. Appl. Environ. Microbiol.. 75(24), 7674–7681 (2009).

168. Ringeisen, BR, Henderson, E, Wu, PK, Pietron, J, Ray, R, Little, B, Biffinger, JC, Jones-Meehan, JM: High power density from a miniature microbial fuel cell using *Shewanella oneidensis* DSP10. Environ. Sci. Technol. 40(8), 2629–2634 (2006).

169. Ringeisen, BR, Ray, R, Little, B: A miniature microbial fuel cell operating with an aerobic anode chamber. J. Power. Sources. 165(2), 591–597 (2007).

170. Nevin, KP, Zhang, P, Franks, AE, Woodard, TL, Lovley, DR: Anaerobes unleashed: aerobic fuel cells of *Geobacter sulfurreducens*. J. Power. Sources. 196(18), 7514–7518 (2011).

Citations

CHAPTER 1

Natalie M. Hughes, Chander Shahi, and Reino Pulkki, "A Review of the Wood Pellet Value Chain, Modern Value/Supply Chain Management Approaches, and Value/Supply Chain Models," Journal of Renewable Energy, vol. 2014, Article ID 654158, 14 pages, 2014. doi:10.1155/2014/654158.

CHAPTER 2

M. A. Islam, M. Hasanuzzaman, N. A. Rahim, A. Nahar, and M. Hosenuzzaman, "Global Renewable Energy-Based Electricity Generation and Smart Grid System for Energy Security," The Scientific World Journal, vol. 2014, Article ID 197136, 13 pages, 2014. doi:10.1155/2014/197136.

CHAPTER 3

Imran Rahman, Pandian M. Vasant, Balbir Singh Mahinder Singh, and M. Abdullah-Al-Wadud, "Swarm Intelligence-Based Smart Energy Allocation Strategy for Charging Stations of Plug-In Hybrid Electric Vehicles," Mathematical Problems in Engineering, Article ID 620425, in press.

CHAPTER 4

Yu-Jun Zheng, Sheng-Yong Chen, Yao Lin, and Wan-Liang Wang, "Bio-Inspired Optimization of Sustainable Energy Systems: A Review," Mathematical Problems in Engineering, vol. 2013, Article ID 354523, 12 pages, 2013. doi:10.1155/2013/354523.

CHAPTER 5

Dan Song, Meirong Su, Jin Yang, and Bin Chen, "Greenhouse Gas Emission Accounting and Management of Low-Carbon Community," The Scientific World Journal, vol. 2012, Article ID 613721, 6 pages, 2012. doi:10.1100/2012/613721.

CHAPTER 6

Nathaniel K. Newlands and Lawrence Townley-Smith, "Biodiesel from Oilseeds in the Canadian Prairies and Supply-Chain Models for Exploring Production Cost Scenarios: A Review," ISRN Agronomy, vol. 2012, Article ID 980621, 11 pages, 2012, doi:10.5402/2012/980621.

CHAPTER 7

Longfei Mao and Wynand S Verwoerd, Selection of Organisms for Systems Biology Study of Microbial Electricity Generation: a Review, doi:10.1186/2251-6832-4-17.

Index